CHALLENGES FOR CHILDREN

Discovering Science Together

Challenges For Children
Discovering Science Together

Barbara Crabtree
illustrations by Laraine Peterson

N.Z. Playcentre Federation
1982

N.Z. Playcentre Federation
P.O. Box 67-085
Auckland

Reprinted 1988

ISBN 0-908609—04—3

Published with the help of
a grant from the McKenzie Trust

cover photograph by Irene Jensen

To
'The sons and daughters of life's longing for itself'
'The Prophet' — Kahlil Gibran

CONTENTS

Foreword	9
Plants	13
Animals	23
Current Electricity	29
Static Electricity	37
Magnetism	41
Light	45
Sound	51
Water	55
Air	61
Making Work Easier	69
Heat	73
Earth Science	81
Appendix: Classification of plants and animals	85
Acknowledgements	87

Foreword

I like having children to do things with — it gives me an excuse to do things I used to feel silly doing alone, like making muslin-men, admiring the big transporters, collecting worms for pets. Doing things with children has taught me to appreciate the simple and the obvious.

Interested adults facilitate children's learning of basic concepts. The aim of this book is to give some guidance to parents concerning possible directions of joint investigations. Every experience has been tried out by pre-schoolers and their parents. Each is related to many daily occurences.

Ideally a child will ask a question.

Together child and adult will ask further questions.

Together they could consult this book for basic information and examples of simple experiences which will aid understanding.

Alternatively, beginning with the experience needs to be seen as simply the beginning. Ask yourself and the children:

- How is this relevant to our lives?
- What other examples relate to this experience?
- Which further experiences/discussions/books would throw further light on this topic?

Neither you nor the child need completely understand what happens. I think all scientists meet moments of truth when, thinking they understand most of a process, they arrive at a point when they can but stand in awe at the intricacy of nature's laws. Many of us frequently stand in awe.

That we *notice* together is sufficient. Our children will grow in understanding beyond our understandings if we lay the foundations now. In 10 or 20 years' time they may still be fitting small pieces of information experienced today into ever-widening and deepening patterns.

The role of the adult in helping children learn scientific concepts

Young children learn through their own actions. Their understanding is limited to their personal interpretation of their perceptions — tasting, touching, smelling, seeing, hearing and bodily sensations provide basic information on which some sense is made of the worlds they encounter.

A child's interpretation of this basic information differs enormously from an adult's. We may label his interpretations as misconceptions but his ideas are not wrong. We must realise that the child:

is making observations
is relating various factors to each other
is connecting different experiences
and *is* working on his own ideas.

The child who makes 'mistakes' is much closer to understanding, than the child who is not prepared to hazard any guess as to reasons and causes until he is sure he can produce the answer adults prefer to hear. I would never correct a child's misconception with 'No that is wrong, this is what it really is'. I would accept his idea but I would remember what he thinks, and expose him to as many experiences as I could which might add to his understanding, and allow him, eventually, to produce his own adequate concept. A concept may take years to develop — there is no hurry. In fact, his thinking can't be accelerated, only encouraged. His concepts are private, and only he can change them. It will take much further action with materials on his part before he will be prepared to change his thinking — and then he will firmly believe he always thought that way.

To illustrate this point, consider a child with little experience with magnets. He feels it necessary to press a magnet hard onto pins, 'so that they stick'. His misconception is obvious: 'my pressing makes them stick on'. In all his previous experiences, pressing does help some things stick together. He will repeat this experience countless times before he realises that the pins hang on without his pressing them; he can bring the magnet very close to the pins and they will 'jump' onto the magnet; he can go 'fishing' with a magnet attached to string, and 'catch' pins, even hang them in a chain — all without any pressing. The only action required on his part is to place magnet and pins in close proximity. His concept of magnetism will be extended when he learns that the magnet can move pins through cardboard, glass, water and wood; and further again if he makes his own magnet, and an electromagnet.

The use of magnets is only a very small part of a young child's scientific learning. Experience in every topic included in this book is still only a fraction of experiences needed in order to construct concepts. Most of a child's learning is through everyday experiences with natural materials, his own body being his first point of reference. Wriggling his fingers and toes, tastings, squeezing and banging, be it in water, sand, food, dough or fingerpaint, all give him basic information about what he can do and how various materials react.

The provision of a safe but stimulating environment, with loving but not smothering supervision will develop his curiosity and confidence. A home which is not too tidy for messy materials, and parents who are not too busy to answer constant questions, and who will draw attention to the everyday environment will extend his interests.

For instance, while cutting codlin moth out of his apple, a mother talked about the apple as the home of animals. What other fruit or plants are the homes for animals?

Adult attitudes which help children develop scientific concepts

1. Encourage curiosity. Answer children's questions honestly and at their level. If you can't, say so and investigate together.
2. Help children to use all of their senses at every relevant opportunity.

   Ask
   - What does it smell like?
   - Can you hear . . . ?
   - Have you tasted this anywhere before?
   - Can you balance yourself on . . . ?

3. Develop children's confidence in solving their own problems.
    - I wonder why . . . ?
    - What if you . . . ?
    - Do you remember what happened when . . . ?

   Restrain your eagerness to tell them why. This demonstrates your learning but probably won't help theirs.

   Explain only at their level when you think they are ready. Under-explain rather than over-explain. Their actions and observations are much more meaningful to them than are your words.

4. Enjoy, and learn with, children.

   For a child, everything is new and different. He discovers electricity *himself*, regardless of Faraday's precedence!

   You don't need to know all the answers. You do need to be prepared to think about underlying concepts, to help children relate their experiences to each other, and to follow-up interests from day to day suggesting further lines of enquiry. In doing so, adults will find themselves learning continuously — and enjoying it.

An 18 month old girl dropped a raisin in her glass of orange. Her hand was small enough to fit into the glass to retrieve the raisin, but large enough to make the orange overflow. The raisin eaten, the phenomenon of the overflowing cup became the absorbing activity. Though her parents disapproved of the mess in the dining area, they did provide a plastic cup in her bath and encouraged further experimentation (eventually to lead towards her concept of floating and sinking).

A 3 year old I know proudly displayed his new shoes and announced 'mine are faster than yours because they're newer'. I accepted the challenge. After winning the first race I suggested we look at the relative length of our strides. He took as big a step as he could and I demonstrated how much further my legs could reach. In our next race, I took steps only the same length as his. He beat me, to my surprise (and I did try). Perhaps Stuart is even more convinced that he is faster than me because his shoes are newer. Or maybe he wasn't really serious when he made that statement. I don't know. However, I hope I reinforced for him that adults are interested in what he has to say; and that there are other factors to consider beyond his first, immediate, perception.

PLANTS

> *Basic Ideas*
>
> There are many different types of plants.
>
> Plants have different parts, each with a different function.
>
> Plants usually grow from seeds, sometimes from stems, sometimes from leaves or roots.
>
> Plants need water, light, air, minerals, warmth and space.

CONTENTS OF THIS CHAPTER

There are many kinds of plants

Some simple plants
Yeast
A mushroom garden
Mouldy old things and real stinkers
Lichens and mosses

Plants have different parts
Leaves
Stems
Roots
Flowers
Fruits and seeds

Things to talk about and do
Make vegetable soup
Find edible flowers
Investigate swollen plant parts

Growing from seeds
Growing beans
Make a wheat man
Sprouting seeds

Growing from stems

Growing from leaves
Miniature gardens

Growing from roots
A hanging carrot basket
Tired vegetables
Bulbs
Easy herbs

What plants need
Cress seed experiences
Make an egg carton garden
Water travels in plants
Compost
On watering cans
Patio-garden containers
Easy saladings
An indoor tomato plant

There are many kinds of plants

Go on a 'collecting' walk with children.
Supply plastic bags in which to collect plants or parts of plants:

leaves,	twigs,
flowers,	bark,
cones,	seed heads,
grasses,	lichen,
green and dry moss.	

During your walk encourage children to think about how tall is the plant — would you call it:
- a tree,
- a shrub,
- grass,
- flower plant,
- fungus.

Does it have one stem or trunk, or several?
Does it stand by itself, or climb, or spread over the ground?
Is it all green?
Are all its leaves the same shape and colour?

Discussion after a collection walk could involve children in finding and matching objects.

'Have you collected a leaf the same as this one?'

'Look at mine; it is pointed at one end and has little points all down its sides.'

'Now can you find a leaf with more than one leaflet all joined together?'

Some simple plants

Yeast

Yeast is a little plant which thrives in warm, wet, sweet places. A tablespoon or sachet of yeast in a small bowl with 300 ml of warm water and a generous half cup of sugar, honey or golden syrup will soon froth up and overflow into another bowl. It has a wonderful smell.

Use yeast to brew ginger beer and to make bread rolls and Easter buns.

A mushroom garden needs tender, loving care.

Damp stable hay packed into a holey plastic bucket (drill holes with a hand drill) which is covered with a few centimetres of soil or compost will usually yield an assortment of toadstools.

NOTE: Toadstools with spongy undersides are **very poisonous!** Many with gills are somewhat poisonous. Only a limited number of fungi are good for eating, though mushrooms are by no means the only ones.

Mouldy old things and real stinkers

Fascinating fungi, and incredible pongs, can be made by keeping a variety of things in covered coffee jars with a centimetre of water to keep them moist.

Try cut pumpkin, bread, bruised fruit, milk, cheese rind, meat and fish trimmings, orange and potato peel, dahlia and cabbage leaves, an egg, an onion.

If you have a fridge, make a double set of jars, and keep the duplicates in the fridge to show that the coldness of the fridge delays the growing things which hasten the decay of dead things.

You may also use a sunny windowsill and a dry set of lidless jars to let some things be preserved by going dry and hard.

Some items could be frozen into blocks of ice for later retrieval.

Lichens and mosses

Use large clear plastic storage jars or small glass-covered aquaria to plant damp-loving mosses, lichens, ferns and houseplants like African violets.

Always collect mosses etc. by cutting out a clod with a knife or pointing trowel, and not by ripping handfuls from banks. Grow them out of direct sunshine, but don't choose a dark corner. Almost all plants need good light.

Plants have different parts

Pull out a small, flowering plant and look at its parts.

How are the leaves arranged on the stem — opposite, in whorls, alternate?

What happens to the leaves in autumn and winter? If they change colour and fall the tree is *'deciduous'*.

If they stay green on the tree in winter, the tree is *'evergreen'*.

Both deciduous and evergreens make new leaves in the spring to use the best weather.

When the chlorophyll goes out of the leaves in the autumn, you can see other colours which the green chlorophyll usually hides.

Use leaves, lichens etc. in collage construction.

Try 3-D montage using plaster-of-paris. If this is mixed with milk it doesn't set so quickly.

Each one has a different job to do.

Leaves

Leaves are for making the plant's food. To do this they need sunlight, water, minerals and air.

Look at leaf points, edges, veins, stalks.

Ask 'Does this plant have different sorts of leaves, or are they all the same?'

Generally the leaves on one plant are of the *same* basic form and shape although size and colour may vary. But many have juvenile and mature leaves which are different e.g. lancewood, silver dollar gum.

Pour mixed plaster-of-paris into shells, small trays or tin-lids — children arrange their collections 'permanently'.

Use leaves for roller-paining

Spread a thin layer of paint on a leaf, place the painted surface on a clean sheet of paper and press down on it gently with a clean piece of cloth.

Carefully remove the leaf — can you see the veins painted on your paper?

Use leaves for printing.

Place a leaf on a white sheet of paper and place another white sheet over the leaf. Using a pencil or crayon rub over the top sheet of paper. You will see an outline of your leaf when you have rubbed over the paper above the leaf.

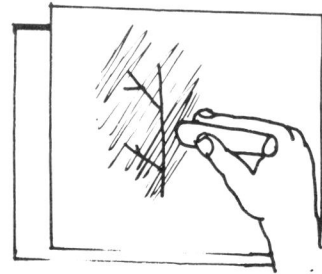

Stems

Plants have different types of stems. Stems are to hold the leaves up to the sun and flowers for pollination. They connect roots to the leaves.

Collect creeping stems (couch), twining (convolvulus), and prickly and hooked stems (blackberry, roses).

Some stems are thin and weak, others are strong, some hollow and some solid.

Other stems remain under-ground.

Roots

Plants have different types of roots. Roots gather water and dissolved minerals, and hold the plant in the ground. Some roots are small and weak, others big and strong.

Use a hand-lens to see the very fine root hairs near the root tips. Or look at a well sprouted seed.

Look at surface roots (in a pine plantation), aerial roots (of philodendron, monsterosa), clinging roots (of ivy).

Watch willow, ivy, carrot, kumara etc. make roots in a jar of water.

Flowers

Flowers are very impressive because they become fruits and seeds. There are male parts and female parts, sometimes in the same flower, sometimes in different flowers. Flowers are for making seeds. When pollen from the male fertilises the egg in the ovary of the female, the seed can start to grow.

Ask about flowers. 'What are they for?'

They look pretty, smell nice and attract insects (bees) for pollination.

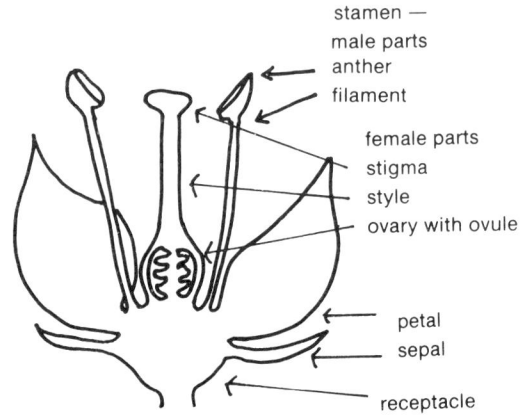

Pull a flower apart and look at different parts.

Talk about sepals, anthers with pollen, the ovary for eggs and seeds.

Look at *different flowers* on one plant. Find closed buds, half-open and fully-open flowers. Ask 'Are flowers all the same on one plant?' Talk about their size, colour, shape and parts.

Talk about different size, shape, colour, fragrance and feel of flowers as children use them in collage, and pressed arrangements, sandsaucers and vases.

Talk about *seasons*. In spring display a branch of flowering black currant or peach with tight buds. Stand in water on the science table, and watch it blossom. In summer bring green then ripe fruits, and in autumn and winter the browning leaves and bare branches. Listen to children's comments and ideas.

'Why are the buds opening and leaves unfolding?'
Because it's spring; summer's coming.'
'Why aren't the leaves on the tree all year?'
Because the tree has a rest in winter.

Fruits and seeds

Fruits are where the plant has its seeds.

Most plants have fruits; not all fruits can be eaten.

Ask children to bring any fruit they can to pre-school: passionfruit, bananas, strawberries, tamarillos, oranges, walnuts, apples. Scrape out all of the fruit or cut it up and place all in a bowl. Eat your fruit salad.

Discuss different fruits. What sort of tree do they grow on? What part of it do you eat? What are the seeds?

- Can you eat them?
- Would they grow?
- What would they grow into?

Taste culinary seeds — sesame, celery, mustard, poppy.

Try some 'vegetable' fruits. Pumpkin, zucchini, cucumber, tomato, corn, peas and beans are vegetable fruits. Look for their seeds. Grasses have grain seeds. Collect wheat, barley and millet (birdseed).

Grow them, or process into flour and bake or make porridge with them.

Discuss **seed dispersal.**
- Pips dispersed by birds (blackberry on beaks)
- Hooks and spines by sheep (biddi-bid)
- Winged seeds by wind (dandelion and sycamore)
- Pods, split and shoot (flax, lupin)
- Human agency — from accidental apple trees to the conversion of forests to grasslands and vice-versa.

Things to talk about and do

Make vegetable soup

Children each bring a carrot, parsnip, onions, parsley, potato.

Provide peelers, vegetable knives and chopping boards.

Encourage the children to do as much peeling and chopping as they can cope with.

Taste the vegetables before and after cooking.

Discuss reasons why we cook food. Talk about the changes in colour, texture, taste, shape, hardness and heat.

Taste before adding salt and seasonings and after.

Find edible flowers

Many flowers can be eaten: violets, scarlet pimpernel, feijoa, nasturtium and dandelion are some of these.

WARNING: Taste only known plants. Many common plants are poisonous.

Investigate swollen plant parts

These are where plants store food and water.
Roots — carrot, parsnip
Stems — Potato, yams, gladioli (underground), ginger, sugarcane, khol rabbi (above ground)
Leaves — bulbs, onions
Fruits and seeds also store food.
Cacti and succulents store water. Grate these or press them through a garlic crusher to see the juices.

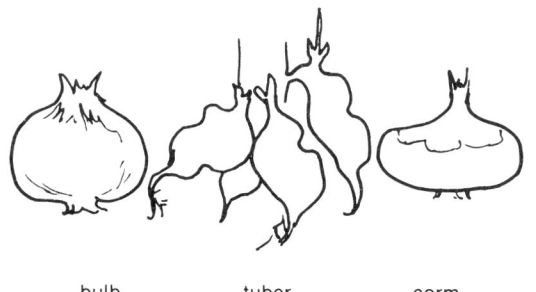

bulb   tuber   corm

Growing from seeds

Plants usually grow from seeds, but sometimes from stems, roots, leaves or tubers.

Talk about seeds. Ask
'What are they?'
'What will they do?'
'What will they grow into?'
'What do they need to make them grow?'

Growing beans

Fill a preserving jar with roughly crumpled newsprint. Place several bean seeds around the jar at different levels.
Keep the paper damp and look at the beans regularly.

Watch the skins split and the roots grow down (always) and the shoots grow up (always).
Put a lid on the jar and turn it upside down.
Watch the roots turn and grow down again while the shoots grow upwards.
Ask why.

Make a wheat man

On a piece of muslin place wheat or cress seeds. Cover with ½ cup of soil or sawdust. Bring the corners of the material together and tie. Sit this on top of a small jar of water making sure that the ends of the muslin are always kept in the water.

Draw a face on your wheatman with marker pens.

Watch his hair grow.
Cut his hair and eat it in a sandwich.

Sprouting seeds

Most large seeds are speedy sprouters — try beans, peas, cress and balsam, marigold and hollyhock seeds.
Plant outside when germinated.
Try planting by the moon for best results. Use an astrology chart (see N.Z. Women's Weekly).

Mung beans can be grown in a rice-cooker. Soak them for a few hours then drain and keep in a warm place.

Rinse through daily to keep damp.
Eat in about three days.

Acorns can be planted half-buried on their sides.

Date pits, avocado stones, whole walnuts and citrus pips will also grow. Make them into ming trees (bonsai) by restricting root growth in small containers.

Growing from stems

Plants sometimes can be grown from stems. Drop cuttings of herbs, pussy willow etc. into a bucket of cold water as you gather them.

Pot them and keep moist under a holey plastic bag supported on sticks, until leaves start to grow.

Some cuttings will take root if stood in a bottle of water — ivy, willows and water fuschia do this very readily. Houseplants such as ivies (Hedera sps.), wandering willie (Tradescantia), fibrous rooted begonias (Begonia maculata) and spider plant (Chlororophytum) are hardy examples.

Growing from leaves

Cacti and succulents — the survivors — are easily grown from leaves. Look for fallen leaves which are sprouting roots and plantlets or detach mature leaves.

Let newly picked leaves lie in the sun for a few days before planting or they may rot.

Miniature gardens can be made in sardine cans. Make drainage holes in the bottom with a hammer and nail then bend the lid back to make a backdrop for your garden.

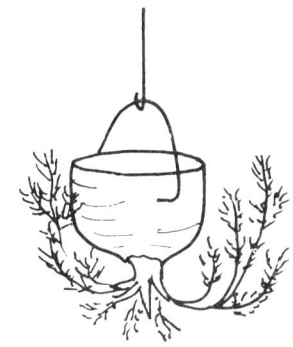

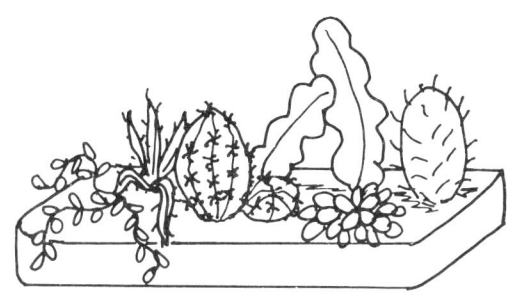

Fibrous begonias and African violets can also be grown from leaves.

Watch shoots grow from the bottom, then turn to grow upwards.

Try growing the root end (without the leaves) of a carrot.

Will it grow?

Tired vegetables

Old, wrinkled or sprouting vegetables can be grown by putting them up to their necks in a jar of water. Any root vegetable will grow in this way and may even eventually seed.

Growing from roots

Carrot tops, kumaras, parsnips and yams can be grown from roots.

Use blotting paper or cotton wool under carrot tops (2 cm thick) and pieces of parsnip to keep them moist, or grow from the whole root by suspending it in water with a little plant food such as Maxi-crop.

Some plants can be grown from fleshy root cuttings — perennial phlox, hollyhocks, oriental poppies.

A hanging carrot-basket

Keep 5cm of the top of a carrot. Scoop out some of the core, leaving all uncut surfaces intact.

Hang the carrot upside-down with string or wire and keep the centre full of water.

Bulbs

Grow a hyacinth bulb, lachenalias, jonquils, grape hyacinths.

Fill a small jar with water and place the hyacinth on top.

Fill a small jar with peat and plant 3 or 4 lachenalias in it.

In a larger pot of peat, plant the jonquil and the grape hyacinth.

Discuss bulbs. Ask 'What are they?'
'What will they grow into?'

Encourage the children to keep them watered during winter and discuss new growth as it occurs.

In general plant spring flowering bulbs in autumn and autumn flowering ones in the spring.

When the leaves die down, after flowering, lift the bulbs and look for baby bulblets. These will grow into flowering sized bulbs over a number of years.

Easy herbs

Herbs are easily grown, from seeds or cuttings. Experiment with a variety of these — ask around for thinnings from neighbours' gardens.

A herb garden can be made in a cut-out tyre, a bath or a washing machine bowl.

Make an egg-carton garden

Put a little peat into egg shells.
Stand these in an egg carton.
Plant a dozen different seeds, or similar seeds, according to your choice.
Label and keep watered.

What plants need

Plants need water, light, air, minerals, warmth and space. These needs have been discussed but they can also be shown by a series of more formal experiments and experiences.

Cress seed experiences

Make sure before you start that there are as few variables as possible.

I once grew cress in egg cups, kept one in the dark and one in the light.

In discussion why the 'dark' one grew pale and weak, a 4 year old said, 'Because that cup's purple and this one's white.' Therefore, use identical containers.

Place cress seeds on kitchen paper or cottonwool in small plastic trays or lids. Keep moist. Place an inverted preserving jar over them.

Try the following, setting up only one at a time.
1. One watered; one kept dry.
2. One in sunlight; one covered with a black cloth.
3. Cover lids with shoe-boxes with a small hole in one end. Watch plants grow towards the hole.
4. One in a cold place; one in a hot water cupboard.
5. One in water; one in soil.

Involve children in the reasoning at every stage.

When you have failures, wonder why. Try again, perhaps at a different time of year. Spring and summer give best results.

Try cactus plants too, which are hardy and last well. Ours is still growing after three years.

Water travels in plants

Colour water in a jar with cochineal.
Stand a white or yellow flower in it — daffodil, carnation, or a cut stalk or celery, having recut the stem under water.

Stand in the sunshine. After only an hour or two the colour will show in the veins of the flower or leaf.

Where did the colour come from? How did it get there? Through the veins.

When celery and flowers are on their plants, the roots take up water from the soil, and it is drawn up through the plant, and after use it passes out through the leaves.

Put some leaves in a jar, and turn it upside down in the sun. Watch water from the leaves condense in the jar.

Compost

Make a compost heap in the autumn when the weeds and grass are growing strongly and need disposing of, leave it to mature during the winter, and when the spring comes there will be a supply of plant food to use in pot plants, patio gardens, or perhaps to grow pumpkins or potatoes.

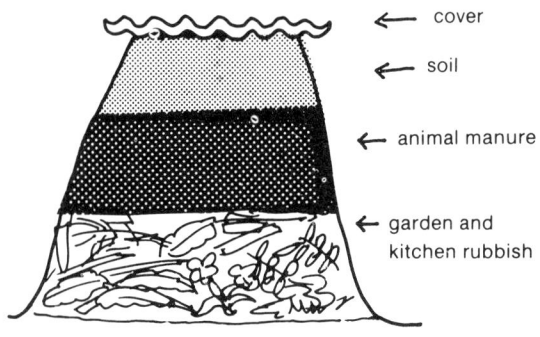

← cover
← soil
← animal manure
← garden and kitchen rubbish

On watering cans

To water seeds, use a fine spray — a used plastic spray bottle will do. Cover the seeds to keep them damp. Pot plants are best watered with a long-spouted plastic watering can, or a teapot. If they get dry over a weekend, dunk the whole pot in a bucket until the bubbles stop rising. Most are best left then until the soil on top gets dry again. Outdoor gardens may be watered with almost any container, can or hose.

Patio garden containers

Defunct bathtubs, washing machine bowls, tumble drier drums, cut down tyres, sinks, tubs, baby baths and halved metal drums can all serve this purpose. Be sure to make drainage holes or whatever you plant, it will drown.

Easy saladings

All the following can be eaten simply cut or grated raw, and are easily grown: beetroot, cabbage, small carrots, apple cucumbers and gherkins, mustard-and-cress, cos and buttercrunch lettuce, bush tomatoes and zucchini. Various wild plants such as chickweed, rauraki (puha), dandelion and nasturtium can also be gathered for salads. Oil and lemon juice shaken together make a fine dressing.

An indoor tomato plant

A climbing tomato grows well when trained around a sunny window. Sometimes water it with Maxicrop, and shake or flick the flowers in the middle of hot days to release the pollen to set the fruit. A mist spray also helps to set the fruit.

ANIMALS

> *Basic Ideas*
>
> An animal is something alive but not a plant. There are many kinds.
>
> Animals eat, grow, reproduce, breathe and move.

CONTENTS OF THIS CHAPTER

Looking at animals

Setting up a terrarium for lowlies

Geckos and skinks

Insects and spiders
Leaf-eaters
Suckers and caterpillars
Carnivorous insects

Mice

Birds

Setting up an aquarium
Pond life
Stream life
Seashore life

Frogs

Goldfish

Guppies

Looking at animals

Most young children are fascinated by animals. Wherever possible, compare animals with what children do and are. Encourage children to ask:
- What is it?
- What does it eat?
    Other animals? Plants?
- Where does it live?
    Does it build a nest?
    Does it need a shelter?
    Does it hibernate in winter?
- How does it move?
    Does it have legs, wings, feelers?
- How does it eat?
- What is its mouth like?
- How does it see? Hear? Feel?
- What are its babies like? Do they hatch from eggs?
- How does it grow? (e.g. life-cycle — does its mother feed it?)

Adults don't need to be able to classify each specimen proudly presented to help children investigate such things as:
  Has it got a hard skeleton on the inside? *(vertebrate)*
  Or on the outside? *(invertebrate)*
  How many legs has it got?
  Has it any wings?
  How many body parts? Insects usually have 3 parts (head, thorax, and abdomen)

We must foster a loving attitude of caring and of being responsible for the wellbeing of the living plants and animals we keep. Find out all about the needs of the things you intend to keep, and make a home ready *before* you acquire them.

Don't keep creatures you are not certain you can look after or that you dislike. If you happen to find something rare, enjoy it where it is, then leave it undisturbed in its own home.

Give 'chance' finds which are brought along temporary homes which meet survival needs, and insist that they be released where they came from at the end of the day.

Most living creatures can be tamed and safely handled, and few serious diseases can be caught from pets. Clean even trivial bites and scratches, however, because infections are always possible, and of course hands should be washed after handling pets. Keep their homes clean and provide amusements as well as food and water, according to the pet's fancies.

Remember to protect living creatures from small children by using child-proof homes — goldfish dislike dough in their water, mice are so easily crushed to death in little fists. Large aquaria lend themselves to the setting up of many kinds of habitat for smaller animals, allow easy vision, and yet give small creatures some measure of protection from unsupervised handling.

Teach children how to handle plants and animals appropriately. Well supervised curiosity can be channelled into a love of nature, a caring for our environment, and a respect for those creatures we share it with.

Setting up a terrarium for lowlies

Use an aquarium, a cracked one may do.

Break open a rotting log with a trowel, put two or three chunks into a plastic bag, and take them back to put in your terrarium. If the log was in a damp spot you should spray water into the terrarium from time to time.

Many creatures may live in the log, including ants, termites, spiders, and horned beetles. If your log contains some **ants,** provide a few crumbs and some sugar water on a piece of sponge for them. To keep the ants from crawling out of the terrarium, spread a layer of vaseline along the upper edge. For an ant colony to thrive you'll need to catch a queen. Watch to see what kinds of insects and other animals come from the log. Some may be eggs when you collect the log and may develop into adults while in the terrarium.

Make a **slater** (woodlice) house. They thrive living in a scooped out raw potato, watermelon rind, or pumpkin shell.

Crickets are delightful, continually changing their skin, size and shape. They come out of hiding if it is warm to sing once they get used to the general noise level. But you have to use a spade to catch them in grass or compost heaps. Cut them a clod of weedy grass and place it on compost. A few piled stones provide chink homes. Sprouting birdseed gives them tender greens.

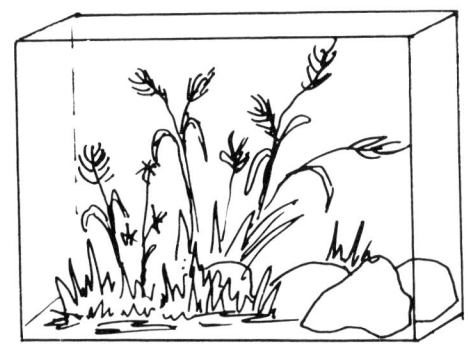

Give **worms, sandhoppers and ground wetas** a lot of compost with a log or clod or turf on top, and a handful of dead leaves. Sow untreated seeds for tender greens.

Exclude light from the sides of the terrarium if you wish these ones to make burrows right up to the edges of the glass. Show worm tunnels in a jar of damp earth covered with black paper.

Geckos and skinks

These need live insects, winged for preference. They hibernate, and can go unfed on occasions. These lizards need drinking water, and enjoy a 50/50 cocktail of honey and water. They prefer a dryish home with narrow crevices to hide in similar to their home territory, be it rocks, split fenceposts, or grassy tussocks.

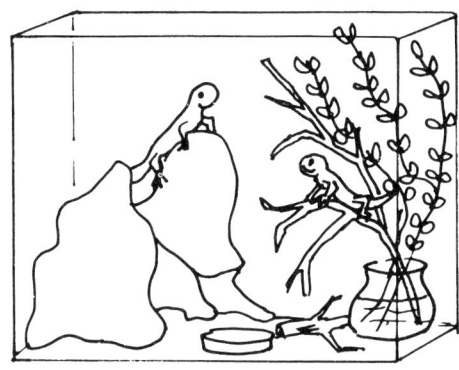

The daytime gecko is the blue-eyed, bright green tree gecko. Naturally, these like fresh twigs. Hammer the ends of the branches, and keep them fresh in a jar of water. Geckos leap, so take care when handling them.

Insects and spiders

All insects must have damp air. A large container filled with earth and kept watered provides this, inside an insect box. Poke twigs of foods into the soggy dirt.

Leafeaters

Stick-insects, wetas, katydids, snails etc. generally like roses, pohutukawa, feijoa, phebalium, manuka and the like. Cut their twigs into a bucket of water, because they need crisp, juicy food. Then push the twigs into the tin of wet earth. Stick-insects do knee-bends if you blow on them.

Suckers and caterpillars

Leafhoppers, aphids and caterpillars must be kept on whatever plant they are found on, and need frequent changes of food. It is often easiest to pot up the host plant.

Keep caterpillars on plant materials placed in a jar of water covered by a disc with a hole in it, lest they drown in the water, or keep only a few caterpillars on a potted host plant: they grow fast, and eat themselves out of house and home, then wander off to pupate. Search on the following plants for caterpillars:

cabbage family — white butterfly
groundsel, cineraria — magpie moth
swanplant — monarch butterfly (they can eat cut pumpkin)
stinging nettle — red and yellow admiral butterflies
beans — silver Y moth
ragwort — cinebar moth
kowhai, lupins — kowhai moth
eucalyptus — emperor gum moth

Carnivorous insects

Mantids, spiders and ground bettles eat other insects or one another. Don't keep them in mixed company, or it won't be mixed very long.

Watch baby spiders hatch from silky egg clusters, but take them outside soon, to fly away on fine threads like kites.

Look for praying mantis egg bases in sheltered spots on the sunny side of garden fences. If the 'lids' along the tops are intact, visit daily until the young pop out. Look for them in the vicinity. They feed on tiny insects such as aphids.

Collecting cans

Punch holes in the plastic lid of a can, and store finds such as snails with the can on its side to prevent carbon dioxide building up. Release or properly house the creatures as soon as possible.

Food bucket

Cut twiggy foodstuffs for small animals directly into a bucket of water to prevent wilting. Insects etc. must have crisp foods.

Mice

Mice have interesting habits. The petshop types come in many colours and patterns, and are friendly and gentle when tamed.

A good mousehouse is a big aquarium with a mesh lid over it. The lid can be made to fasten with magnetic catches, or a suitcase strap. Put 10 cm of dry, undusty litter in the bottom, coarse sawdust, leafmould, chainsaw chip, pine-needles or earth for instance — mice like to burrow in it.

Furnish the mousehouse with a ventilated bedroom — a large unglazed pottery plant pot with a doorway chipped out is fine, or a wooden box. Provide an insulating material for bedding, like hay or apple wrappers, which don't collect dampness the way newsprint does. Mice arrange a neat round nest in the middle of a great heap of bedding. Have the bedroom in one corner, and the mice will then use the diagonally opposite corner for their toilet.

Clean this corner regularly and the whole environment should remain sweet for several weeks. Female mice are odourless, but males make potent stinks if they are frightened, excited, or busy staking claims to territory.

Put their food and water near the bedroom. It may be conveniently kept on a mezzanine floor above the bedroom if space is limited. Food must be uncooked. Buy mixed grain, or pellets (which they don't much like), or mix linseed, hulled oats and millets. They can use budgie seed. Raw bone and insects provide minerals, and greens — chickweed, groundsill, rauraki, dandelion, and grass seedheads, particularly fescues — provide vitamins. Greens can be kept fresh in a jar of water with holes punched through the lid for stems. A drip bottle for water saves fouling of water, mice dump all sorts of junk in water containers.

Mice are most active at night, but do come out and feed and play during the day. Give them a treadwheel to exercise on, placed where nothing can tangle in it. Other playthings enjoyed are branches to climb, cones and logs, compost, bones or fruit tree prunings to gnaw (they must constantly wear down their teeth), and cardboard tubes for tunnels, and things which roll. Don't give them a loaf of bread. It looks cute, but mice quickly become obese, and have the usual problems fat people are prone to.

If you intend to keep more than one mouse, remember males will fight practically to the death, but females are peaceable. They breed better than monthly if allowed, and the average litter is eight, so remove the male if the female seems pregnant — it shows. Give her plenty of bedding, water and extra food. Don't disturb the nest and babies until they open their eyes at two weeks, and begin to venture out. Wean them and segregate the sexes at 28 to 30 days. Sex them by matching with their parents' appendages.

Catch and lift them briefly by the tail base, and have children hold mice in a container or *on flat hands*, avoiding the use of the thumbs, for mice are fragile. Hurt or sick ones sometimes respond to warmth, dark and isolation, but can be humanely killed with ether on cottonwool in a closed jar if necessary. Their normal lifespan is two years.

Birds

Budgies and canaries are easily kept cage birds. Zebra finches also do well. Finches like a private corner, such as a purchased cane nest. Give them dried grasses and such, but no thready fabric or wool scraps, for nesting materials. You may choose to grow their greens.

It is possible to keep bantams or even hens or ducks!

Bird and mouse greens

Bird (and mouse) greens include silver beet, chickweed, rauraki (puha), chewings fescue, grated carrot, dandelion, grass seedheads, shepherds's purse, clovers, scarlet pimpernel etc. They are a daily essential.

Setting up an aquarium

Use as big an aquarium as you can accommodate, because the amount of oxygen in the water relates to the size of the surface area. A large goldfish bowl will do. Tap water needs to be left to stand before use to get rid of the chlorine — pond or rain water is better.

Use sand or smooth pebbles, washed if they are dusty, to cover the bottom generously, and place water plants weighted with stones in the corners where they can root. They will provide greens and help with oxygenation and waste disposal. Put in some large stones and hollow things for hiding-places.

When the bottom needs cleaning, siphon off the sludge, then replace the lost water with fresh. Top up between cleanings too. A sheet of brown paper to pour on prevents stirring up the bottom.

Keep the aquarium in strong light. Pond snails help to control green algae. Clean the glass with steel wool or washed fleece wool if slime becomes annoying.

Pond life

Use mud, plants and water from the pond — take buckets for these. Set up the aquarium, let it settle, then go back for backswimmers, water boatmen, tadpoles, cyclops, daphnia, and whatever else you can strain out with a pantyhose leg stretched over a wire hoop or can with both ends removed.

Most live near the mud, and prefer warm, shallow water. Feed these on fine fish food, decaying leaves, and dead insects hooked on fuse wire for easy removal.

Some of the tiny pond animals make fascinating subjects for a few minutes with a microscope or a simple hand lens or magnifying box.

Stream life

Stream life is a bit more difficult because of the higher oxygen requirements of the running water habitat. Keep stream animals only long enough to have a look at, then replace them.

Seashore life

This needs special care. Look at it, and then leave it right where you found it. Take home dead material. Seaweed, shells, and driftwood stitched to hessian can make lovely decorative screens. Hung in netting or a hammock they can make room dividers.

Frogs

Frog spawn may be found resembling half-cooked sago among shallow pondweeds. The eggs will hatch, and the tadpoles thrive in a 'pond' aquarium.

The tadpoles initially eat decaying vegetation, and later progress to decaying meat. They cannot metamorphose unless iodine is present, but usually they grow their legs etc. normally. At that time they also change to air breathers, so add a beach for them, or a sloping rock or log.

When they have all their legs, reduce the pond to a bog. Leave a puddle, and add a clod with marginal grasses and plants growing on it. Have a suitable hiding-place, like a piece of punga. The aquarium will now need a mesh lid.

Frogs must have live food, and will accept almost anything which wiggles. Worms can be gathered by wagging a spade shoved into a lawn, or woodlice found under junk. Moths can be caught at dusk, and flies are easily trapped in a meat-baited jar with a cut-off funnel in the top.

Collect the flies in a plastic bag by tilting the funnel, and drop them in, bag and all.

I have a particular fondness for the Australian tree frog found on the West Coast and elsewhere, and kept a pair for 12 years before freeing them in a fernery. They visited on rainy nights after that. They hibernate conveniently in the winter, and become quite tame. They need mossy branches suited to their curious lifestyle, and the usual damp conditions frogs enjoy.

Goldfish

Goldfish like to have a bolt-hole, so provide a tunnel, perhaps a length of field tile, which is non-collapsible. Don't use shells or concrete, as they dislike limey water. They need plenty of space, so be content to start with a couple of small fish, and watch them grow.

Feed goldfish on fish-food from the pet shop, and give them live mosquito larvae, daphnia, etc., as often as possible. These can be grown for the purpose. They also enjoy aphids, caterpillars, or a scrap of liver to suck. Liver must be removed within a couple of hours because it decays very quickly. Golfish will accept crushed breakfast cereal in an emergency, but avoid bread, which swells inside them.

Fish should never be overfed. See that there are no leftovers from the daily food, and remember that an occasional missed day won't matter if they have weeds to munch.

Some commercial fish foods leave a slick of oil in the water, which causes the fish to gulp for air at the surface, and to make bubbles. Remove the oil by dragging a length of toilet paper across the surface and gathering up the oil, before the fish suffocate.

Guppies

Guppies need a heated aquarium, but are colourful little fish, and have live babies in profusion. The excess can be fed to goldfish, or bigger tropicals.

CURRENT ELECTRICITY

Basic Ideas

Current electricity needs a circuit to travel in.

Electrical energy can travel through some things and not others.

Electricity has a magnetic effect.

CONTENTS OF THIS CHAPTER

Safety first

Materials

Electricity needs a circuit
A simple circuit
A switch in a simple circuit
A motor in a simple circuit
Make a colour-wheel

How electricity travels
Insulators and conductors

Electricity makes magnetism
Make an electro-magnet

Bonus experiences
Dismantling a flat battery
How electricity is produced
Circuits
How a bulb works
Make traffic lights
Make a simple energy model

Safety first

Stress with children the danger of playing with any household electrical circuits — sockets, plugs, switches, worn leads, fallen wires can be lethal!

Materials

- 1½ volt dry cell batteries (No. 6) are easiest to use. These are the tall cylinders with 2 terminals on top (either screw-capped or the clip connection). Any small torch battery can have wire soldered on, but these need protection from constant mishandling by preschoolers. Batteries up to 6 volts are perfectly safe.

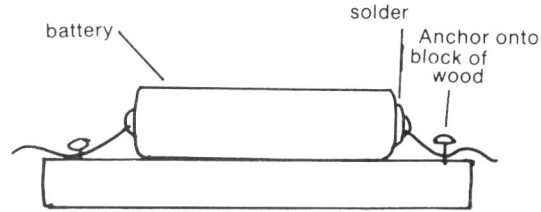

- Bulb-holders come from some electrical supply shops.
- Torch bulbs (screw-in) are readily available.
- Wire — insulated single or multi-strand. Pester auto-electrical repair shops for gizmos with lots of wire all over. Cut into short lengths and strip the ends for a few centimetres. Wires need bare ends at connections to battery, bulb, switches, and motors.

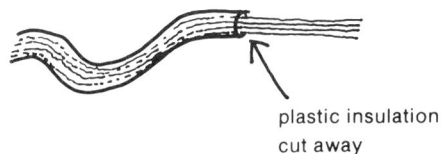

It helps children to cope with screwing the terminal cap onto the wire if the wire is twisted into a loop and put on the terminal so that the end loops in a clockwise direction:

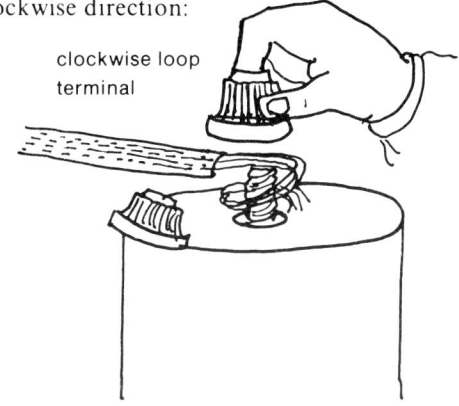

clockwise loop terminal

- A motor from toy and model shops, for use in model boats. They are cheap, and come in three sizes — (I use the smallest, 120). These will last much longer at preschool if mounted on a wood-block.

- A home-made switch. (See a switch in a simple circuit.)

Teach children to always switch off after using the equipment to avoid wasting the battery.

Electricity needs a circuit

Connect one bulb in a circuit with one battery:

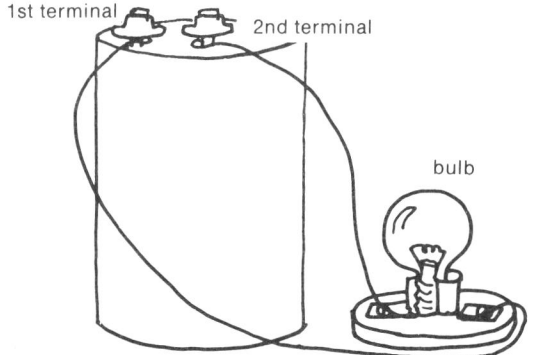

1st terminal
2nd terminal
bulb

When children have problems getting the light to 'go', remind them that the electricity needs a whole circuit to travel in. Ask 'Where is the circuit broken? Which metal bits are not touching?'

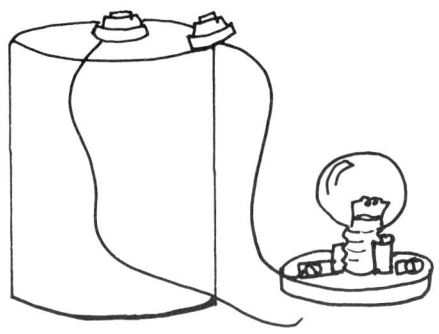

Trace the circuit with your finger — positive terminal to bulb holder, through the filament and out the other side of the bulb, and back through the other wire to the negative terminal. 'What needs to be joined to make the bulb light up?'

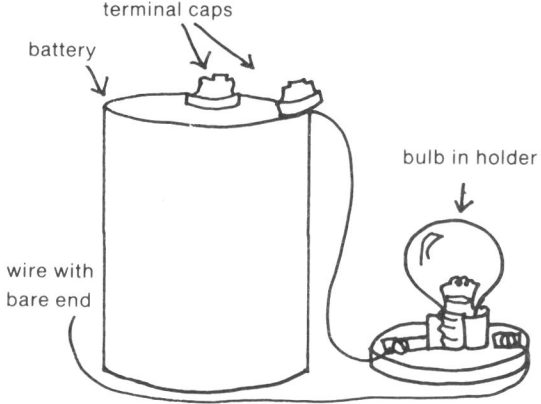

terminal caps
battery
bulb in holder
wire with bare end

How can we turn the light on and off?
Electricity is being used to make light.

Warning: Never leave a bulb lit when not in use.

A switch in a simple circuit

Connect a simple switch into the circuit.

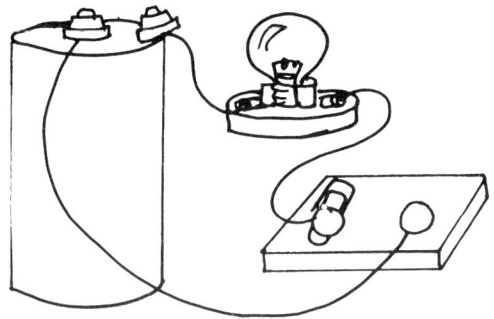

Adding a switch to the circuit.

The switch makes it easy to turn the current on and off.

When the switch is closed, electricity runs through the paper clip. We say the circuit is *'made'*.

When the switch is open, electricity can't run from one drawing-pin to the other. We say the circuit is *'broken'*.

Check that the ends of the wires are stripped of plastic insulation, and that the bulb is screwed into the holder firmly.

Warning: Always turn the switch off when not in use.

A motor in a simple circuit. (Children's favourite.)

Make a circuit of one battery, one motor, one switch.

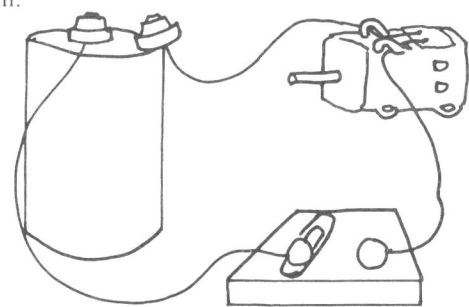

Feel the vibrations as the motor turns.

Make a colour-wheel

Use any stiff paper in any shape. Pierce the shape with a pin and push onto the motor shaft. Try small circles and squares, and vary the position for the shaft — try the centre of the shape, then a corner or edge, and note the difference when it turns. Use a blank shape, and while it spins gently hold a mini-marker against it.

Use any pattern and colour combinations. Talk about colours mixing, and how speed changes the appearance of the shapes.

Warning: Turn off the switch when not in use.

How electricity travels.

Insulators and conductors

We can sort insulators from conductors.

Insulators are things which electricity *cannot* go through.

Conductors are things which electricity *can* go through.

Make a circuit of one battery, one bulb and one switch, without the paper clip.

Have on hand an assortment of small objects to place across the drawing-pins, ensuring that you have more conductors than insulators, e.g.:

keys	paper-clip
cork	hair-clip
wood	button
pencil	inner seal
rubber-band	tin-foil
milk-bottle tops	wire (bare of insulation)
wire with insulation	nails — shiny, galvanised, and rusty

Before trying each item ask 'Do you think this will make the bulb glow?'

Try it.

When the bulb lights up this means the circuit is complete.

Does the electricity run through keys?

When you put paper across the drawing-pins, does the bulb light? Why not?

Experiment with all the objects.

Sort the objects into two piles.

Pile one which electricity 'likes'. We call these objects *conductors* — and pile two which electricity 'doesn't like'. We call these objects *insulators*.

This is an obvious example of logic being developed with basic scientific concepts (insulation and conduction).

Turn the switch off when not in use.

Relate these experiments to everyday insulation and conduction.

e.g. porcelain insulators on power-poles

plastic insulations around wires children are using

heavy insulations around telephone and power poles.

• People are good conductors of electricity. This is *NOT GOOD* for people who touch electricity! ***STRESS*** to never play with plugs, switches, fallen wires, etc.

Electricity makes magnetism

Make an electro-magnet

It is probably better to do this experiment after becoming very familiar with ordinary magnets.

Materials:

1 10 cm nail, and 3 metres of wire. (Shorter lengths may be joined, but remember to bare all the ends first.)

Connect 2 or 3 1½V batteries in series, i.e. in a straight line with the negative terminal always joined to the positive on the next battery in line; Or use a 6V battery.

Coil the wire around the nail as many times as possible. Coils on top of coils are fine. Join one end of the wire to an end terminal, the other to a switch. Connect the switch to the other empty terminal.

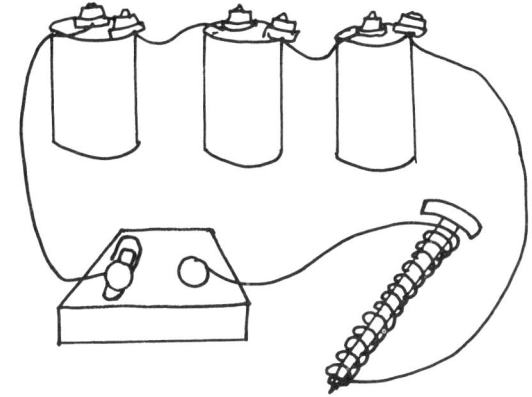

Test the end of the nail as a magnet. Will it pick up pins?

Close the switch. Now do you have a magnet?

Test many items and sort into two piles:

One pile attracted by the 'electro-magnet'.
One pile not attracted by the 'electro-magnet'.

Is this sorting the same as for ordinary magnets?

Feel the nail. Is it warm?

Electricity needs to push to get through the nail; this is work, and it makes heat.

Turn off the switch. Is the nail still a magnet? I wonder why? Did the electricity change the nail?

An induction coil from a radio speaker also works well.

Drop the electro-magnet. Now is it as good a magnet?

Explanation: A single wire carrying electricity has a small magnetic field around it. By putting many coils around the nail we simply increase this magnetic field sufficiently to use it. The nail, as the core, is magnetized, that is, its atoms are aligned in a regular pattern, and thus exert a magnetic force-field. It stays magnetised for some time after the current is switched off, unless a sharp drop shakes the molecules in the nail back into random directions.

Bonus experiences

Dismantling a flat battery

This is safe, but messy. Wear old clothes.

Materials:
Flat battery,
Pocket knife — use with care,
Lots of newspaper.

Remove the outer cardboard case of the battery.
   Now you can see what the outer (negative) terminal is attached to.
   The outer case is made of zinc.
   Cut through the zinc case.
   You'll discover a black sticky paste, then black powder.
   In the centre is the positive terminal — a carbon rod.

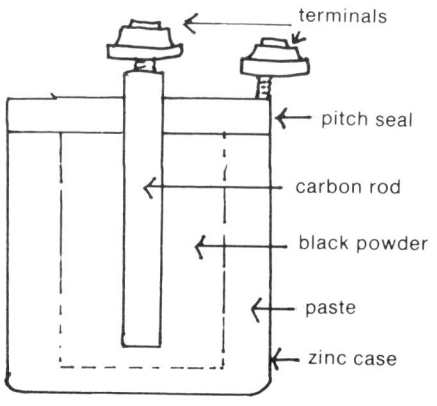

How electricity is produced   (mainly for parents).

Electric current equals flowing electrons, i.e. free electrons flowing in one direction in a wire from one atom to the next (like a train shunting). All materials consist of atoms — some atoms hold on to the electrons much more firmly than others. Carbon holds on firmly to electrons, and in fact, readily collects any spare ones around. Zinc lets its electrons go readily.

The paste inside the zinc case loosens some of the electrons in the zinc atoms. These drift very slowly towards the carbon because the black powder is a barrier. However, when a wire connects the zinc to the carbon the electrons race atom to atom along it, from the zinc to the carbon. Nature always takes the easy path, and copper wire is easy for electrons to travel in, i.e., is a good conductor. As it loses electrons the zinc is changing. It changes gradually to zinc oxide which is soft and the battery begins to leak.

Circuits

Examine a torch. Pull it apart to trace the path of the current. If the outer case is metal, it may be part of the path.
   How does the switch work?
   Why do some torches use more than one battery? (If yours does, the terminals must be joined positive to negative.)
   Try your bulb in a circuit with two batteries — connect them this way:

Is the bulb brighter with two batteries?

Look at the bulb-holder — inside and underneath.

Trace the path of the current. One screw which holds the wire connects via a plate underneath to the bottom of the bulb. The other screw attaches its wire to the socket into which the bulb screws. Thus the circuit is complete.

How a bulb works

Materials:
Torch bulb in bulb-holder (socket)
Magnifying glass.

Using the magnifying glass, look right inside the bulb. You can see 2 wires supporting a tiny, coiled wire. Connect the bulb in a circuit, use the magnifying glass again to see the coiled *filament* light up.

Holding the bottom of the bulb unit on top to the other terminal, you'll find you can touch the free wire end anywhere on the brass base of the bulb and the bulb will light.

Make traffic lights

Use a three-way switch for this.

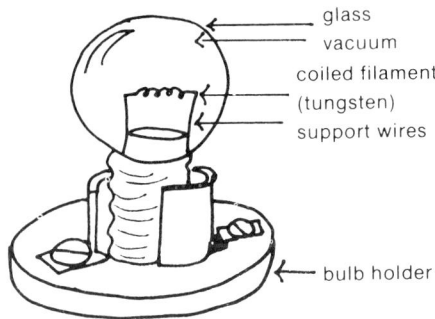

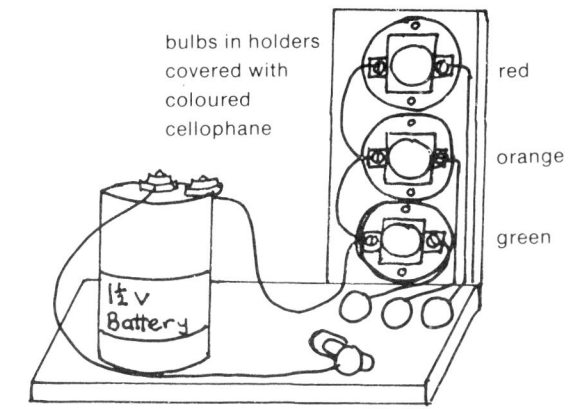

The finer the wire, the harder the electricity has to work to push through it. As it pushes, electricity makes light. The wire is made of tungsten which is very tough, and the tiny coiled wire lasts well.

All the air is removed from the bulb, then the glass is sealed so that the filament does not burn away. (In a vacuum it can't burn.)

Unscrew the bulb and remove from holder.
Connect one wire to a battery terminal.

My eight year old constructed this, my six year old wired it up, and my two year old loved lighting each traffic light on request.

Make a simple energy model

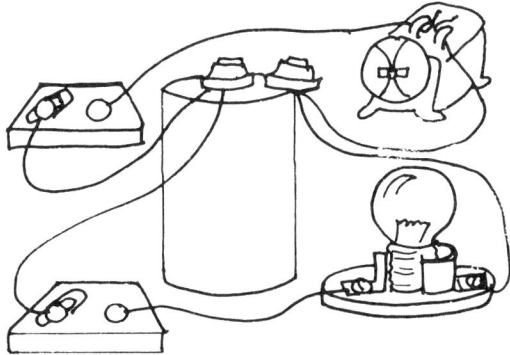

This model shows how electrical energy from the battery can be converted to:

- heat and light energy in the bulb

- sound and motion (vibration) and some heat energy (it gets warm) in the motor, which can be used to do

- work (another form of energy) as in turning colour-wheel.

What happens when both switches are closed at the same time?

STATIC ELECTRICITY

> *Basic Ideas*
> Static electricity is produced by rubbing.

CONTENTS OF THIS CHAPTER

Materials

Make your own electricity
Tear tissue paper
Blow up balloons

Make your own lightning storm

Materials

- A fine dry day. Any dampness in the air 'absorbs' free electrons, making these experiences impossible.
- Ball-point pen or plastic rod.
- Wool, fur, silk, or nylon — either scraps or worn garments.
- Combs — plastic or nylon.
- Cotton thread — short length.
- Flat, clear plastic containers (cheese segment boxes are good, or icecream containers will do).
- Tissue paper.

Read 'How does a battery work.' (Bonus experiences, Current Electricity).

Make your own static electricity

Take a 3cm square of plastic, and watch it fall when you let go.
 Now rub the plastic quickly on wool, fur, silk or nylon.
 Let it go.
 You let go, but the plastic doesn't.

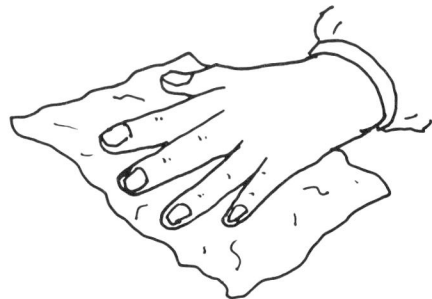

Your rubbing gave it an electric charge.
This makes it cling to you.

Tear tissue paper into small bits.
Place them inside the plastic container. Replace the lid and rub it very fast with fabric (wool, hair or fur). Watch the bits of paper — they jump dramatically to the lid.

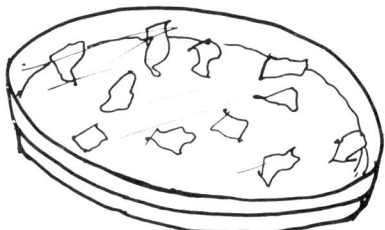

Rubbing built up an electric charge on the lid. This charge attracted the bits of paper, just as the plastic was attracted to our hand.

We could build a charge on the comb by rubbing it with fabric or on our hair. Try.

Some children's hair charges much more easily than others. Test the charged comb for attraction to cotton, thread, bits of tissue paper and other people's hair.

Blow up a balloon and tie with string.

Will it stick to a wall?
Build a charge by rubbing with fabric.
Does it stick to the wall now?

Try it near bits of tissue paper or the hairs on an arm.

Can you relate these results to experience with magnets where like ends repulse and unlike ends attract?

There are positive and negative charges of static electricity. Taking away electrons leaves the balloon positively-charged, as when rubbed with wool; adding electrons leaves the balloon negatively-charged, as when rubbed with silk.

Like charges repulse;
Unlike charges attract.

Make your own 'lightning storm'

Blow up a balloon.
 Hide in a dark place, e.g. under a blanket.
 Rub the balloon fast many times with fabric (20-30) to build up the static charge.
 Lick your finger, then watch carefully as you bring your finger close to the charged balloon.

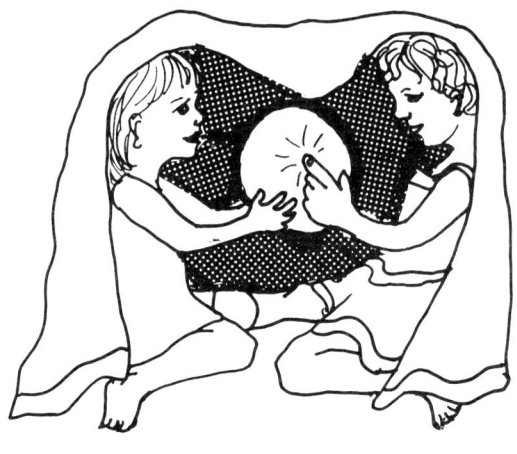

Do you see, or hear, the tiny spark jump to your finger? Only if conditions are perfect will you see sparks, but you should hear them.

Relate the spark-to-wet finger to *never touching switches,* and especially bathroom switches, with wet hands. (Water on the hands quickly attracts electricity; you can't let go, and burns and shock result.)

Relate the experience to *lightning storms.* An electrical charge builds up on water drops rubbing together in a cloud. When differently charged clouds come close, the charge jumps from one to the other.

Lightning heats and rapidly expands the air. We hear this disturbance as thunder.

We see the lightning first, and the thunder is heard later, because light travels faster than sound.

Relate this to seeing or hearing sparks of electricity jumping when clothes come out of a *tumble dryer,* (especially synthetic and woollen garments dried together). And pulling clothes off, (especially, again, a combination of wool and synthetic). The warmth from the dryer or your body has made them really dry, so there is no moisture to absorb the charge. The rubbing together builds up charges.

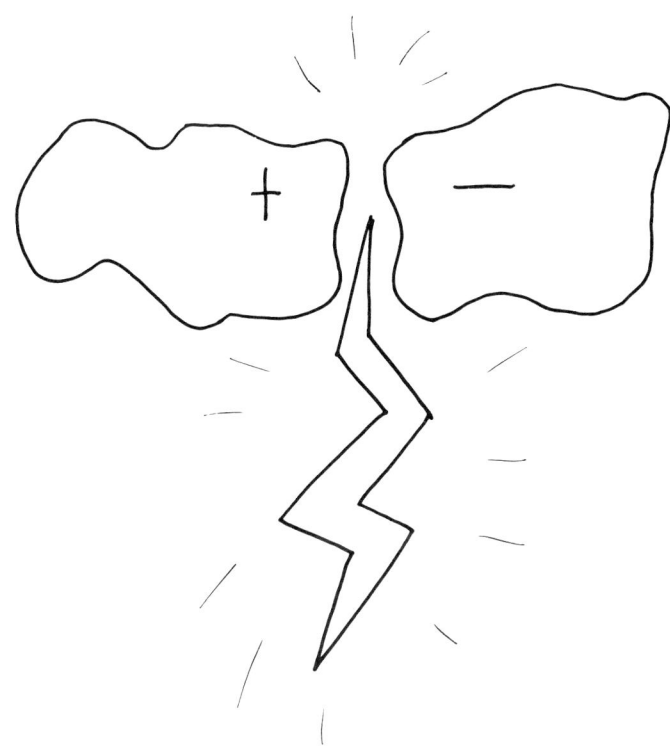

MAGNETISM

Basic Ideas

Magnets attract some things.

Magnets attract through many materials.

Most magnets are strongest at each end.

Magnets have north and south poles.

Magnets make some other things into magnets.

CONTENTS OF THIS CHAPTER

Materials

Magnets attract some things
Play a sorting game
Play a fishing game

Magnets attract through some things
Cardboard
Glass
Water

Magnets are strongest at each end
Make the magnetic force-field visible

Each end acts differently

Bonus experiences
Make your own magnet
Make your own compass

Materials

- A selection of magnets — horseshoes, bar, others.

  Sources for these include: old telephones, motors, water meters, electrical repair shops, school supplies. (Very cheap ones are generally unsatisfactory.)

  Note: A horseshoe magnet should always have a 'keeper' across its poles when not in use. Any iron or steel object will do. This retains its magnetic strength. Don't drop a magnet. This destroys its magnetism.

- A wide variety of materials, e.g. paper-clips, nails (both rusty and shiny), bolts, keys, tin-foil, milk bottle tops, cork, hair-clips, steel wool, buttons, plastic pegs, ballpoint pens, wood, paper, polystyrene, glass, brass, coins.

  Use these things to investigate the powers of your magnets.

Magnets attract some things

Play a sorting game

Have an exploratory session sorting items into piles — a 'yes' pile, items of which the magnet attracts, and a 'no' pile which the magnet doesn't attract. Containers labelled 'yes' and 'no' are helpful.

Notice that children at first feel the need to push the magnet onto the object. Encourage them to hold the magnet very close to see if the item is attracted to it.

Ask, 'Do you think the magnet will attract this thing? Try. Which pile does that go in?'

Ask children to observe the things in the 'yes' pile. Do they look alike? In what way are they alike? You may be surprised at how much children know.

Play a fishing-game

Children make their own fish out of cardboard — say 10 cm long, and coloured as they like. Push a paper-clip on for a mouth, or use a staple.

The fishing-rod is a magnet tied to string and a small stick to hold on to.

Place the fish on the floor, or in a basin, then the children go fishing.

This game is lots of fun!

If children have their name on a fish, it helps them to learn to recognise their own name, and other names.

You could put colours on the fish for colour recognition.

Magnets attract through some things

Cardboard

Put a pin on top of a sheet of cardboard.

Hold a magnet underneath, and try to move the pin around.

Does the magnet work through the cardboard?

Try the magnet under, and the pin on top of, a chair, a table, an icecream container and books. Make a magnet maze!

Glass

Wonder if a magnet works through glass.

Put the pin inside a glass and see if you can draw it to the top, using the magnet.

Water

Will the magnet work through water?

With the pin in the glass, cover it with water.

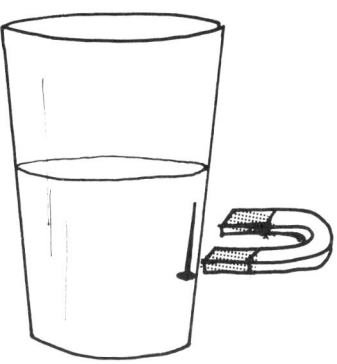

Can you take the pin out of the water without getting your fingers wet?

Most magnets are strongest at each end

Which parts of the magnet attract things most strongly?

Using horseshoe or bar magnets, try the ends and try the centres.

Use pins, cut up steel wool, and chains.

Make the magnetic field visible.

Sprinkle tiny tacks, iron-sand or fine steel shavings onto a shoe-box lid. (Beware: if at pre-school, cover the lid with glad-wrap to save spillings. Iron filings on magnets are very difficult to remove.)

Hold a strong magnet underneath and gently tap the lid.

The iron will make a pattern along the lines of force, with most of them at the ends of the magnet.

Try different magnets. Is the shape of the force-field the same?

Each end acts differently

Suspend a bar magnet (A).

Bring the end of another bar magnet (B) towards one end.

Do they repel or attract each other?

Try the other end of magnet B.

Having marked the north on both magnets, experiment with them.

Can you touch the north of one to the north of the other?

South to south?

Do they push each other away? (*repulse*).

Whan happens when you put unlike ends together, south to north, do they pull towards each other? (*attract*).

Allow plenty of time for experimenting.

Relate the behaviour of magnets to experience with current and static electricity.

> *Like charges repulse.*
> *Unlike charges attract.*

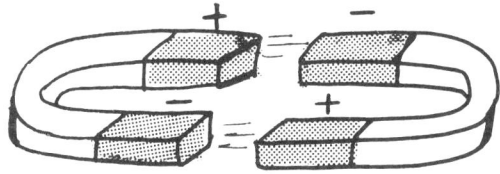

Bonus experiences

Make your own magnet

Using a darning needle and a strong magnet, stroke the needle across one end of the magnet in one direction.

Count aloud about 30 strokes.

Test your needle as a magnet using the 'yes' and 'no' piles as in A.

Drop your needle several times. Is it still a magnet?

This is why we need to be careful not to jar a magnet — even a strong one can be spoiled by dropping.

Could you make a magnet of a pencil?

Make your own compass

A darning needle magnet may remind children of a compass. A compass is simply a magnet; because it always points north it can be used for direction finding. If you can support it on top of water, or by string, you have your own compass.

Extension: Look for uses of magnets in everyday life: door catches on cupboards, fridges and freezers, in motors, on some torches, magnetic holders on fridges, on some can-openers.

LIGHT

Basic Ideas

Without light it is dark.

Shadows are made when some light is blocked.

White light is made up of all the colours of the rainbow.

Light shines through different materials differently.

Light can be bent.

Different materials reflect light differently.

CONTENTS OF THIS CHAPTER

Materials

Without light it is dark
Make a peephole (shadow) box

Shadows
Making shadows — torch
Making shadows — sunshine
Day and night

White light
Make your own rainbow
Make white light

Testing transparency and opaqueness
Cellophane paper experiences

Light can be bent
Bend a straw
Water droplet magnifier

Good reflectors and poor reflectors

Materials

- Thick blanket
- small objects
- square block
- ruler
- dish
- circle of cardboard
- string
- magnifying glass
- newspaper
- sponge
- transparent and opaque materials
- shoeboxes
- torch or circuit with bulb
- ball
- pins
- small mirror
- protractor
- punch
- straw
- waxed paper
- medicine dropper
- reflective materials

Without light it is dark

With light, we use our eyes to see.
Ask children, 'What makes us able to see?'
Hold hands with a friend. Shut your eyes tightly — can you see your friend? Can our eyes see in the dark?
Drape a heavy blanket over a small group of children. Can you see each other with your eyes open?
Now, what do we need to see?

Eyes and light

So where does dark come from?

Make a peephole (shadow) box:

In a shoebox draw pictures, or paste upright objects — shapes — cotton-reels — woodshavings.
Cut a small peephole in one end.
Secure the lid on the box.
As each child looks through the peephole ask what they can see. Why can't we see much?
Because it's *dark*.
Connect a light into the top of the box.
Use either a bulb-holder — wire — battery circuit. The battery can stand inside the box, cellotape bulb and wires to the inside of the box lid and connect to a switch on the outside of the lid.

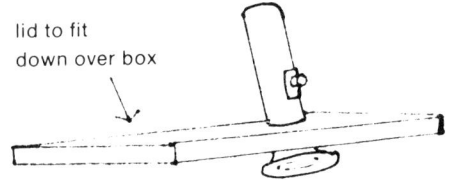

lid to fit down over box

or push a small torch through the lid, keeping the switch on the outside.

Let the children turn the light on and view through the peephole. Ask: Why can we see so much more now?

The viewing is even more spectacular and fun if done under a heavy blanket.

Shadows

Shadows are made when some light is blocked. Shadow play is fun.

Making a shadows — torch

Shine a torch at a wall.

Make funny shapes with your hand between the light and the wall.

Try different shapes to see what shadows they make.

Does a square make a square shadow whichever way you hold it?

Does a circle?

Can you make shadows of different lengths with a ruler?

What makes the shadows?

Your hand, or the shape, stops light.
Where light doesn't shine it is dark.

Making shadows — sunshine

Outside, on a sunny day, play shadow games.

Can you jump on your own shadow? . . . your friend's shadow?

Is your shadow just as long in the morning, at lunch time, and in late afternoon?

Light can't shine through your body, and this makes your shadow.

Fill a matchbox with plasticine and stand a pencil in it.

Put the box on a sunny window-sill, look at the shadow at different times and see how it moves.

Day and night

Day and night are made by shadows.

At night-time half the earth is in its own shadow. Our light comes from the sun.

As our earth spins around, sometimes we can see the sun — daytime — and sometimes we can't see the sun — night-time.

Make a model using a torch for the sun, and a ball (orange) for the earth.

Push a pin into the orange (pretend it's you).

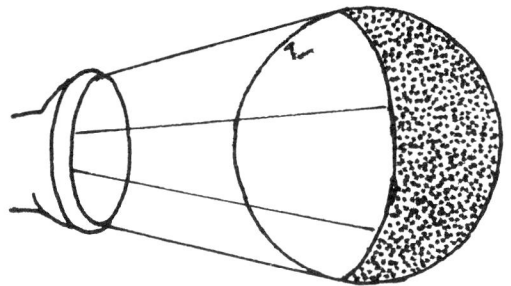

See how you move from day to night to day... as you turn the round ball on its axis.

If you can go high up hills overlooking the sea's horizon, or if you look from a plane, you can see the curve of the earth. That the earth is round is a difficult concept for young children.

White light

White light is made up of all the colours of the rainbow.

Look for rainbows in the sky when it is raining and sunny, in waterfalls, in the spray of a garden hose when the sun is shining behind you, and in soap bubbles.

Make your own rainbow:

Put a few centimetres of water in a small dish.

Hold a small mirror in the water so that it catches strong sunlight.

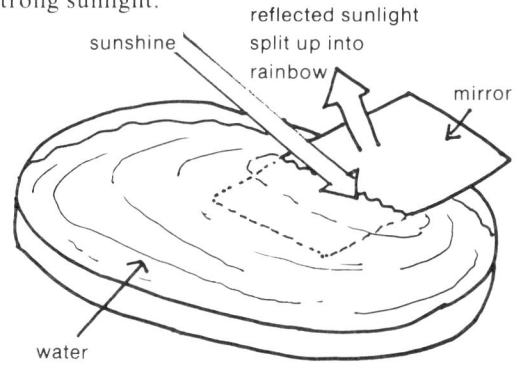

A rainbow will appear on the wall.

What makes the rainbow?

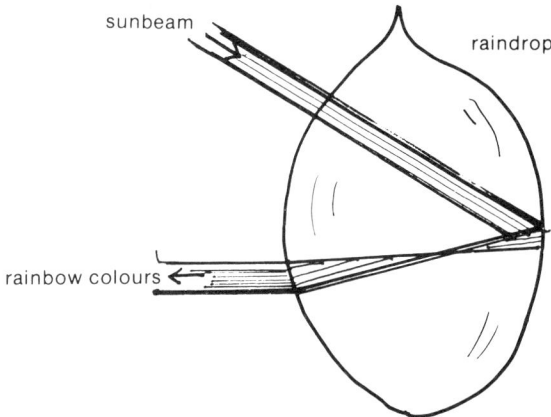

The water splits the white light into its different colours.

We use the mirror to reflect the colours back where we can see them. With a real rainbow, the sun shines through raindrops and splits into colours, which reflect back to us.

Make white light

Divide a circle of cardboard into 7 equal segments.

Paint each segment a colour of the rainbow, in the order red, orange, yellow, green, blue, indigo and violet.

Make 2 holes on either side of the centre about 1 cm apart.

Thread string (1 metre long) through these holes and tie. Wind spinner up by holding ends of loop with first fingers, flip to begin, and then use full arm motion. When fully wound, pull hands apart and let the string go slack.

Watch the colours as your circle spins.

You will no longer see the colours you painted, but a mixture. When exactly the right colours are used we see a white circle, but it will probably be a dirty grey mix, because of impure colours.

Try your rainbow-colour circle on your electric motor (see Current Electricity). It will spin even faster than your hand spinner.

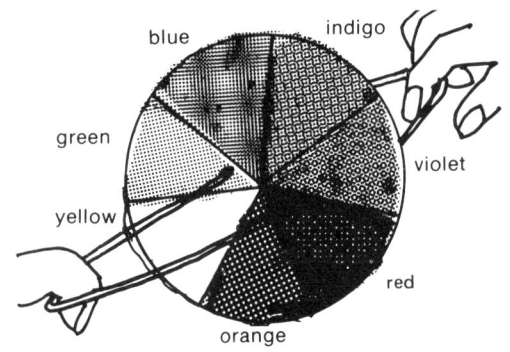

Testing transparency and opaqueness

Materials:

A lidless shoebox with a large hole cut in the bottom.
Torch.
A selection of materials to test for transparency:
    cellophane paper — clear and different colours
    plastic
    waxed paper
    white sheeting
    dark material
    woollen jersey
    sheet of perspex
    white paper
    black paper
    bottle of milk
    jar of water

Large heavy blanket over table.

Adult and children put their heads under the table in the dark, the children being opposite the adult who has the torch.

Shine the torch through the hole in the box towards the children.
    'Can you see the light?'
    'Does light shine through air?'
Hold the plastic across the hole and test.
Then the clear cellophane paper.
    We call these materials *transparent*.
Transparent means light can shine right through.
    Try waxed paper. The wax spreads some of the light. Light can't shine straight through.
    Try white sheeting, then patterned sheeting.
    Try dark material, white paper, black paper and the bottle of milk.

Materials that absorb light are called *opaque*.
    'Why have we got a heavy dark blanket over us?'
    It keeps out the sunlight so we can see the torchlight better.
    The blanket is *opaque*.

Cellophane paper experiences

Try coloured cellophane paper, at first red, or green, or blue, then red and blue together, green and yellow together. These could also be held up to a window to allow sunlight to shine through them.
    Talk about mixing of colours.
    Make cardboard 'glasses' and put coloured cellophane 'lenses' in them. Do they make any difference to the colours you see?
    Paint or draw a traffic signal: look at it through different coloured cellophane (red, orange, green).
    Are the colours the same each time?

Light can be bent

Bend a straw

Stand a straw in a glass of water and look at it through the side of the glass. It looks 'bent'. The water bends the light on the way to your eyes, and makes the straw look bent.

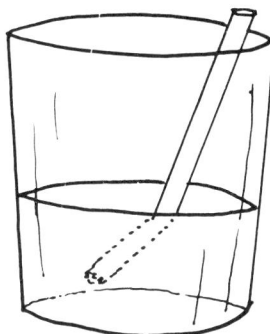
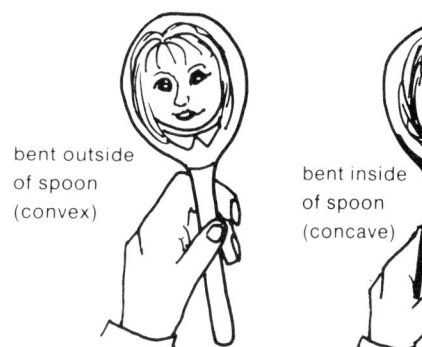

Water-droplet magnifier

Place waxed paper over a newspaper. Using a medicine dropper put one drop of water on the waxed paper. Do the letters on the newspaper look different? Bigger or smaller?

Add further drops of water, until the water is no longer curved. Do the letters still look bigger?

What makes them look bigger?

The curved water.

Use a sponge to absorb the water, and try again.

Compare the drop of water with a magnifying glass. Look at the shape of the magnifying glass — does it bulge outwards in the middle? The curve of the water, and of the magnifying glass, bends the light and makes things look bigger.

Good reflectors and poor reflectors

Look at yourself in
- a mirror
- in a potlid
- in a baking sheet and
- in a spoon

In the spoon do you look different in the curved side (the concave side), or in the curved out side (convex side)?

Can you see yourself better in some things than in others?

Try reflecting a torch in these objects.

Catch the reflection on a white cardboard screen, or on the ceiling.

What makes the light?

When the torch is off, does the mirror reflect on the screen?

The torch makes the light, the mirror *reflects* it.

Try some poor reflectors —
          plastic covered books,
          white card;
and some non-reflectors —
          black cardboard,
          towel.

Black colours absorb all the light which hits them.

Red colours absorb all the colours except red.

Orange colours absorb all the colours except orange (red and yellow).

Can you think of a big reflector in the sky?

The *moon* makes no light of its own.
It reflects the sun's light.

At night when we are on the dark side of the earth, the sun's light bounces off the moon into our eyes.

SOUND

Basic Ideas

We hear sounds through our ears.

Noise is made by something moving. Where there is sound there is vibration.

Different sounds are made by different sizes of vibrating objects.

Sound travels through materials.

CONTENTS OF THIS CHAPTER

Materials

Listening
A sound-matching game
Make tape-recordings
A cardboard cone
Animal ears

Sensing vibrations
Yoghurt pot telephone
A speaking tube
Church bell spoons
Further experiences

Investigating pitch
Twang a ruler
Hang different-sized dowels
Blow across bottles
Make pan pipes

Investigate musical instruments

Sound travels through materials

Materials

- Containers with lids
- tape-recorder
- yoghurt pots
- matchsticks
- hosepipe
- comb
- papers
- wood blocks
- bottles
- corrugated card
- plasticine
- alarm clock
- conventional musical instruments
- items to make shakers
- drum
- string
- funnels
- spoons
- tissue paper
- tin trays
- rulers
- sets of metal kitchen utensils
- straws
- bike pump
- electric motor
- musical box
- nails

Listening

Ask children to be very still and quiet.
 'What can you hear?'
Go on a 'listening walk'. Adults and children will hear things long taken for granted. Find words to match the sounds: the *chirp* of the bird; *scrunch* of footsteps on stones; *screech* of tyres on roads. Hear loud sounds; soft sounds; pleasant sounds; unpleasant sounds.

Make a sound-matching game

Place various small items in different small tins, e.g. rice, split peas, pebbles, buttons.
 Make 2 tins of each item.
 Children can then match the sounds.

Make tape-recordings of common sounds

cat mewing	baby crying
dog barking	children laughing
paper scrunching	chair scraping

Children guess what the sounds are. Keep them simple at first.
 Children love making their own sounds to play back — they may mime the original action, with great delight.

While listening to noises ask children to shut their eyes.

'Can you still hear the noises?'

Now, open your eyes, cover your ears tightly with your hands.

'Can you hear just the same?'

We hear with our ears.

Talk about noisy places — road works (trucks, bulldozers), town, the kitchen at 5 p.m.

Talk about quiet places — bush, the bedroom at night, at church.

Make a cardboard cone

Use it to catch sounds (*ear trumpet*), and to throw sounds (*megaphone*). The cone from machine knitting wool works well.

Animal ears

Animal ears are shaped to catch noises.

Talk about rabbit and mouse ears which are shaped like a funnel. Elephants have enormous ears.

What about
> a praying mantis with 'ears' on its elbows,
> a frog with 'ears' on its head,
> a fish with 'ears' along the line in its body?

Sensing vibrations

Put seeds, or tiny bits of paper, on a drum.

As you strike the drum, watch the seeds jump.

The skin of the drum *vibrates,* sending the noise into the air.

Yoghurt pot telephones

Make a pair of 'yoghurt pot telephones'. Beer cans work well too.

The strings need to be pulled taut.

Two sets for two children saves confusion of who's to listen, who's to talk.

Ask 'When you talk into the 'telephone', how does your voice get to your friend?'

Along the string.

With both children listening rub your fingers along the string.

Ask children, 'Can you hear the vibrations my fingers make?, When I talk to you, how does my voice reach you?, What might my voice travel in?'

Children will understand that a voice travels in the air only if they have grasped the concept of air being all around us.

A speaking tube

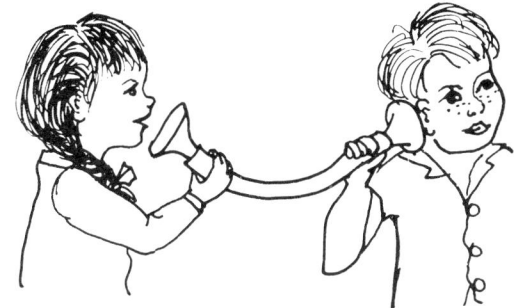

Use two funnels and a length of hose for a 'speaking tube'. This is easier than the yoghurt pot telephone — it need not be taut and it can bend round corners.

Church bell spoons

Tie spoons together in a group on a short length of string.

Hold the string to your ear and jangle the spoons on the other end.
 Do they sound different if you don't hold the string to your ear?
 What does the sound travel in?
 If the string is held to your ear it travels through the string.
 If the string is not held to your ear it travels through the air.

Further experiences

Experience an alarm-clock and a music-box.
 When they are run down, do they make any noise?
 Can you feel any vibrations?
 Wind them up. Listen! Feel! Vibrations make sounds.

Using the 30cm pieces of garden hose to whisper in, ask, 'Is there anything in the hose?'
 Yes, air.
 'What did the sound of your name travel in?'
 The air.

Gently place your hand over your voice-box.
 Say, 'aaaahh . . .'
 Can you feel your voice-box vibrating?

Hum through a comb covered in tissue-paper.
 Pull finger along teeth of a comb.

Shake seeds, sand, pebbles in squeezy-bottles.

Crumple cellophane, newspaper, onion skins.

Bang a tin tray.
 Bang two blocks of wood together.

Investigating pitch

Twang a ruler on the edge of a table

Try a long length over the edge.
Try a short length similarly.

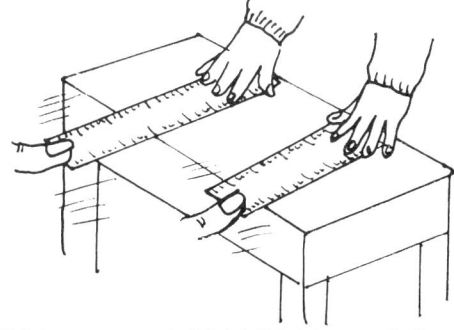

Which one sounds high? Which sounds low?

Hang different sized dowels, bamboo or nails, on a taut string

Tap with a rod.
'Can you play a tune?'
If you could get different lengths of narrow pipes cut, hang them similarly.
Tap different-sized items of various sets to listen for different pitches of sound, e.g.,

> nesting steel bowls
> heavy pot-lids set,
> nesting steel measuring cups.

Blow across the tops of identical bottles with varying amounts of water in them:

'How many notes can you make?'
Try blowing the air of a bike-pump across the tops.

Make pan pipes:

Materials:

2 rectangles strong corrugated cardboard, 5 cm x 2½ cm approximately,
5 plastic or paper straws cut to different lengths.

Glue the straws into the corrugated card.
 Plug the bottom ends of the straws with plasticine.
 Alter the pitch by pushing the plugs up the straws.

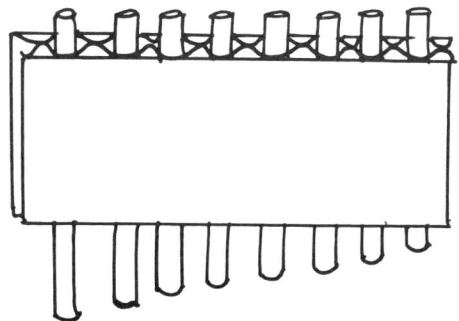

To play, blow across the tops of the straws, or try blowing with a bike-pump, and note any difference.

Investigate musical instruments

The autoharp shows the difference in sound (pitch) produced by different lengths of string very clearly. Also look at an upright piano with strings exposed, or a grand piano with top open.

With a recorder investigate different sounds produced by having fingers over different holes, i.e., different lengths of air vibrating.

Sound can be produced by striking (drum, piano, xylophone, castanets) —plucking (autoharp, ukelele, guitar) —blowing (recorder, flute, trombone) —rubbing (violin, rezorezo) — or shaking (maracca).

Sound travels through materials

Wind up a loud alarm clock — place it on a table and listen to it ticking.
 'Put your ear on the table — can you still hear it ticking? Does it sound different?'

Noise travels through the wood of the table.

Wrap the clock in a blanket. Again listen for the ticking.
 'Does it sound softer than travelling in air and in the table?'

Try the same experiments using tapping of fingers, a music-box, an electric motor.
 Discuss the 'church-bells' and 'yoghurt telephone'.
 Does sound travel better (is it louder) through air, or through string?

Sound travels very well through water — this is why we need to be very quiet when we go fishing. And it is why it hurts fish in aquaria when you tap the glass, and why you shouldn't bang stones together while swimming under-water. **Loud sounds can damage the ears.**

WATER

Basic Ideas

Water is all around us. All living things need water to survive.

Some things absorb water, others repel water.

Water supports floating objects.

Water can be liquid, solid or gas.

Water has weight.

Some things dissolve in water, others do not.

CONTENTS OF THIS CHAPTER

Materials

Water is all around us
Living things need water

Testing materials for absorbency
Wetting and drying

Some things float
Floating and sinking experiences

Water can be liquid, solid or gas
Ice and snow
Condensation
The water cycle

Water has weight
Water pressure
Make a siphon

Testing soluble and insoluble substances

Learning about water concepts with food

Materials

- Saucers
- Medicine dropper
- spray bottles
- objects which float & sink
- potlid
- salt
- hose
- waterwheel
- candle
- seeds
- absorbent & non-absorbent materials
- hair-drier
- electric frypan
- egg
- ice
- soluble & insoluble materials

Water is all around us

Natural water is in lakes, rivers, the sea and in puddles.
Where does tap water come from?
Discuss how it is treated. What with? What for? Where?
Discuss what we use water for — drinking, cooking, cleaning, swimming in etc.
Look around for water — puddles, tanks, creeks, toilet systems.

Living things need water.
Water is needed by all living things.

Count how many cups of liquid you drink each day. Remember that cordial and milk are mostly water. Don't forget milk on weetbix or in custards.

Watch cows, cats, dogs, and birds drinking. Discuss how they drink. Do they suck, or lap the water?
Can animals live without water?
Can we live without water?

Grow seeds or carrot tops on cotton-wool or kitchen paper, using three saucers.
Water two lots.
Don't water the other lot. Does this one grow?
Stop watering one of the two you began by watering. Does this one continue to grow?
Can plants live without water?

Testing materials for absorbency

Materials:
Different cloths — cotton, wool, hankies, sponge, wood (polished and unpolished), cotton wool, paper tissues, stones, e.g., coal and pumice, dry clay, plastic.

Spray bottle or medicine dropper.

Sort materials into those which *quickly* absorb water and those which don't.

Drape a square of dry cloth over a child's hand.
Spray with water, using a spray bottle.
Repeat with a square of plastic on the other hand.
Discuss which materials would be best for raincoats. Why?

Wetting and drying

Talk about material absorbing and repelling water.
Which materials dry out quickest, the good absorbers or the poor absorbers?

Leave various wet materials in the sun or hang on the line.
Hasten the drying process with a hair-dryer.
Have similar articles in the sun and in shade —which ones dry faster?

Children love to 'paint' concrete or wood with brushes and water.
Does it dry quicker in the sun and wind?
Talk about the concept of moving air (wind), especially if it is warm, evaporating water.

Try dipping a finger in water and draw on a blackboard.
Now blow with your mouth on part of the picture.

Test wind direction by holding up a sucked thumb.
The windy side dries quickest and feels chilled.

Some things float

Children will be gaining experience whenever they play with objects in water. However, adult-child discussion is essential for consolidation and extension of children's developing logic in this area.

Collect an assortment of objects for children to experiment with in the water-trough, e.g.:
pieces of wood, corks,
a corked bottle, empty, tin lids,
a corked bottle, full, stones,
of sand, a ping-pong ball,
plastic, a golf ball,
plasticine, toy plastic key,
pumice, metal key,
green wood, rocks.

Encourage sorting objects into 2 piles — those which sink and those which float.

Floating and sinking experiences

When an object is placed in water it displaces some water e.g., when you get into the bath, the water level rises.

If the object is lighter than the weight of the displaced water, it floats, e.g., a cork float.

If the object is heavier than the weight of the displaced water, it sinks, e.g., a lead sinker.

The weight of water displaced by the object depends on both the *weight* and the *shape* of the object.

Put an empty plastic bottle on water — it floats. Fill a similar bottle with sand and place on water — it sinks. Both bottles are the same shape, i.e., flotation depends on weight.

Screw up a milk bottle top into as tight a wad as possible.
Place on water — it sinks.
Place a flat milk bottle top on water — it floats.
Both tops are the same weight, i.e., flotation depends on shape.

Heavy boats, made of concrete, iron and steel and carrying heavy loads can float because of their shape.

Experience upthrust of water on objects

Push a block of dry wood into water.
   Feel the push of the water against the wood.
   Let go under water and the wood 'pops' to the surface.

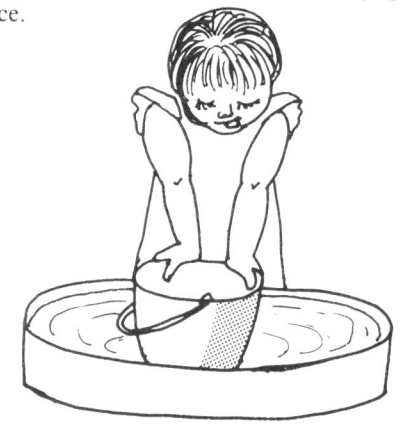

Try to push an empty plastic bucket into water.
   The water works hard against you.
   Fill the bucket with water and there's no problem sinking it.

Water can be liquid, solid or gas

Ice and snow

Make ice, look at it, taste it, smell it. In cold areas it's easy to collect ice and/or snow and watch this melt.

Try freezing water with food colouring in it. Big blocks are beautiful — leave one on the grass and watch it melt. Watch children's actions and listen to their comments.

Notice different rate of ice melting in sun and shade. Put one saucer of ice in sunshine and the second saucer of ice in shade.

Put some ice in an electric frying-pan (check that the cord is out of any passageway).

Listen to the children's ideas about melting, freezing, water, bubbles, steam, fog.

Condensation

Hold a cold saucepan lid over steam and collect droplets of water.
   Bring in language such as evaporation, condensation, hot vapour (steam).
   Ask 'Where else do you see condensation?' (On car windscreens, on windows on a cold morning, on milk bottles fresh from the fridge.)

All air has some water in it.
   Warm air holds more water than cold air.
   When warm air is cooled (as on the potlid, or the cold glass of car or window) it has to drop some of its water. This is the condensation we see. Set up a condensation jar, by putting coloured ice into a lidded jar. Ask children 'Where could the moisture on the outside of the jar come from?'

Water droplets in clouds, fog or steam are so small that they are suspended in air. When droplets come together in cold air they become too heavy to float and fall down as raindrops.

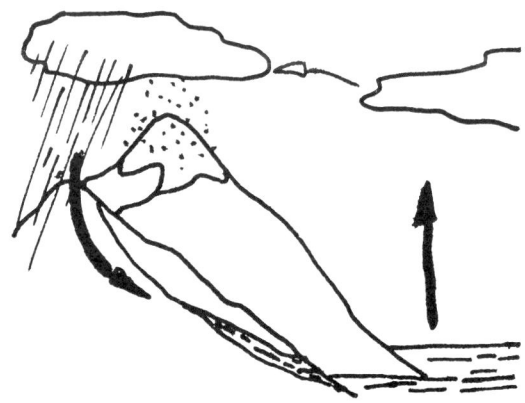

The water cycle

The heat of the sun evaporates water from the sea, lakes, and rivers, and from the ground, plants and animals. This forms water vapour which is suspended in the air. When air is cooled, usually by being forced to rise, some of this water vapour will condense.

As more and more water condenses the drops get bigger, until they are heavy enough to fall to earth as rain, or snow or hail, if air temperatures are cold enough.

The rain water may evaporate again; or fall to the earth to be used by plants and animals. It runs into streams and rivers and back to the ocean to continue the cycle.

Water has weight

Use a glass of water to begin discussion on the weight of water.
'Is water heavy?'

Ask a child to hold a bucket.
'Will it get heavy if it's filled with water?'
Fill the bucket gradually.

Knowing that water has weight allows better understanding of buoyancy and water pressure.

Water pressure

Using hammer and nail make holes all round the sides of a tin can at the same level.

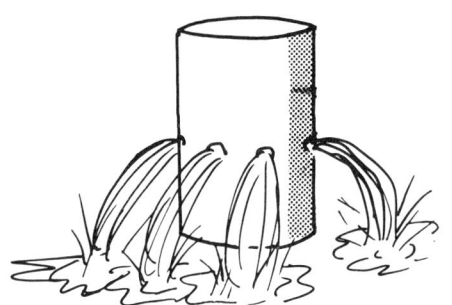

Fill the can with water and notice that the water shoots out the same distance from all the holes.

To show that depth affects water pressure, make 3 holes at different heights in a tall tin.

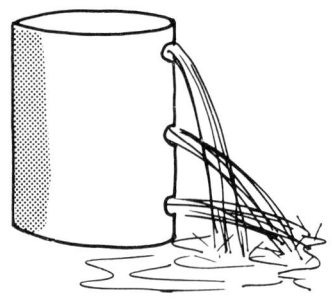

Pouring water in at the top, notice from which hole the water shoots out the furthest.

When water is placed in a container, its pressure is greatest at the bottom. The taller the container, the greater will be the pressure at the bottom.

Make a siphon

To siphon from a water trough into a bucket, first fill a length of hose with water, by either immersing both ends in the water, or by sucking out all the air from one end, while the other end is in the water.

Put one end of the hose in the bucket on the ground while the other remains in the trough.

When the bucket is full, lift it higher than the level of water in the trough. The water will siphon back into the trough.

Does your pre-school or home have a tyre or container which fills up with rain water?

Try a siphon to empty it out.

Remember all you need is a hose full of water and the free end lower then the end in the water which needs emptying.

How a siphon works

Water flows out of the hose because of gravity, i.e., the weight of the water.

Air pressure pushes water into the hose at the higher end to replace that flowing out the lower end. This pressure is sufficient to push water 'uphill' the short distance necessary, so long as no air bubbles enter the hose to push against incoming water.

Testing Soluble and Insoluble substances

Use glasses or plastic containers half full of water.

Put a teaspoon of sugar into one container and stir it.

Ask 'What happens to the sugar? Can you see it? Taste the water — is the sugar still in it?

The sugar is dissolved in the water.

What happens when you stir other things in the water? e.g.:
- salt,
- seeds,
- cocoa,
- chalk powder,
- clay,
- flour,
- sand,
- orange-drink crystals,
- flavoured jelly crystals,
- gelatine.

Try dissolving sugar, drink crystals, salt, gelatine and jelly crystals in paired glasses of hot and cold water.

Ask 'Which dissolves things more quickly, hot or cold water?'

Use language such as dissolving, solution, and gel.

Place a piece of clean cloth over an empty glass.

Pour the solutions through this filter.

Ask 'Is there any sugar, salt, flour, etc., left on the cloth?'

If there is, this is the amount of substance which didn't dissolve.

Try shaking oil in a capped container of water.

Does it seem to dissolve?

Allow it to stand for a while.

Sort materials into 2 groups — those that dissolve in water and those which do not.

Having investigated dissolving of jelly crystals in hot and cold water, set the jelly. When it is set it can be cut, and wobbled. Will it now dissolve again in water (hot or cold)?

Learning about water concepts with food

Mix *jelly,* then set it. Note the change from solid (crystals) to liquid to gel.

Freeze fruit juice ice-blocks.
Try filling a plastic container right to the top, put a lid on, and see what happens when it freezes. (Water expands when it freezes).

Cook rice. Talk about water absorption and expansion.

Pop corn. High temperature changes the water in the corn to steam, which explodes the corn. Steam is powerful — remember the steamdriven water-wheel?

Make pikelets in an electric frying-pan. Adults keep your hands behind your backs and give only encouragement.

Make soup. Children cut up the vegetables, dissolve flavourings, stir and watch alphabet-mix expand. Then eat.

Make *milk-shakes:* by shaking with an airtight top;
 by blowing through straws;
 using an egg beater.

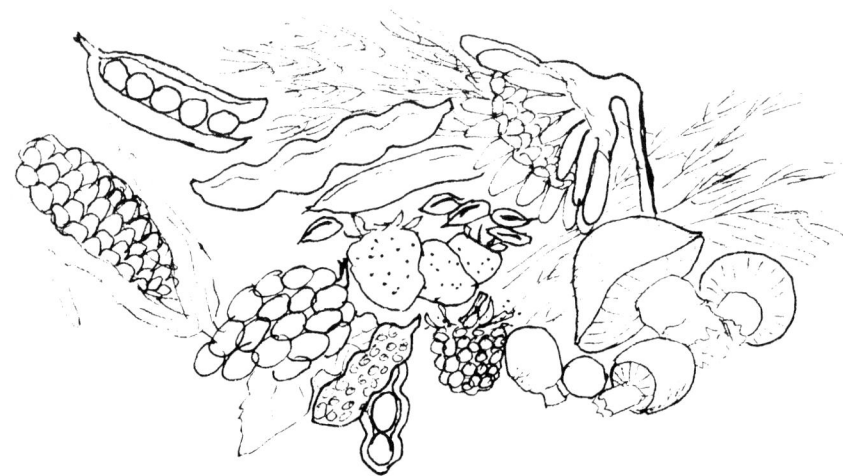

AIR

Basic Ideas

Air is all around us.

We breathe in air.

Air has pressure.

Warm air expands.

Young children struggle with the concept of air being real. They cannot see, touch, taste, smell or hear it, unless it is moving or contained. They become intrigued by experiences which will help them understand the basic concept.

CONTENTS OF THIS CHAPTER

Air is all around us
Feeling air
Seeing air
Pushing air

We breathe in air
Prepare a blowing bag
Hose blows

Air has pressure
Pushing air against things
Pushing air against liquids

Warm air expands
Blowing up a balloon with a bottle

Bonus experience
A tetrahedral kite from milk straws

Air is all around us

Feeling air

You will need empty plastic bottles, shampoo or similar sized.

Ask 'Are those bottles empty? Point the bottle at your chin and squeeze it. Do you feel something? Is something coming out of the bottle? What? Does the bottle look empty? *Is* it really empty?'

The bottle is full of air. Offer each child a tissue, or a feather.

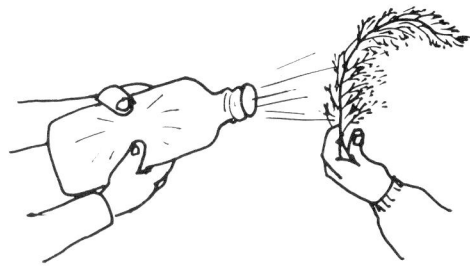

Point the bottle at the tissue as you squeeze it. If the bottle is really empty nothing can come out when you squeeze it. What is pushing the tissue? What was in your bottle?

Children may relate to wi<u>n</u>d. Wind is moving air. We are making bottled wind!

Seeing air

Place the bottle under water.
Ask: 'What is being pushed out of the bottle?'

Repeat with a medicine dropper.
What is squeezed out when you squeeze the rubber?
  If you squeeze it under water you can see bubbles of air floating to the top.
  When you let the rubber go, water is pushed into the medicine dropper to take the place of the air.

Pushing air

Is an empty can really empty?
Using a nail and hammer, put one hole in the bottom of an empty can.

Ask: 'Is this can empty?'
Invert the can in the water-trough — if it's not deep enough, use a bucket. As the can is pushed into the water, ask the child to hold a finger or a tissue just above the hole.
Ask 'What can you feel? Is something being pushed out of the can? Is the can really empty?'

Immerse a sponge, blotter, soil, bread, or brick in water. What do you notice?
Bubbles of *air* floating to the top. Air is lighter than water.

These experiences will need to be repeated many times before the concept of air being everywhere is grasped — and still a 4 year old will tell you that 'bubbles of water are coming out of the bottle'.

We breathe in air

What do we breathe in and out?
Hold a tissue near your face as you breathe in and out. What happens to the tissue? What is making the tissue move?
Try with your lips closed.
With your lips open.

There are two ways for the air to go in and out of our bodies, through our mouth and our nose.

Go for a run around a building. Come back and put the tissue near your mouth. What difference do you notice?

Prepare a Blowing Bag

Cellotape a straw securely into the top of an airtight plastic bag.

STRESS SAFETY — plastic bags are very dangerous to play with because they can block your mouth and nose so you can't breathe. Always use bags too small to fit over a child's head.

Blow into the straw, then hold your finger over the end.

What is in the bag now?
Squeeze the bag and feel what's coming out of the straw (air; wind = moving air).

Hose Blows

Cut short lengths of hose.
Ask children: 'What is inside this hose?'
    They'll probably say: 'nothing'.
Now blow in one end with a child's hand near the other end.
    'What is coming out?'
    Wind, or moving air.
    Blow through the hose into water. See the bubbles of air. Sometimes add detergent to the water, also food-colouring or tempera paint.

Blow through a straw onto your hand, or under water.

Air has pressure

Pushing air against people

Stretch out your hand and swing it quickly around your body.
'What can you feel?'
    We cannot see air, but we can feel it when we are moving.

Now hold onto a piece of paper while you swing your arm.
The paper bends as it is pushed through air.

Watch tree leaves blowing in the wind.
Give children sheets of newspaper, outside.
Each child holds a sheet in front of himself.
'Will the paper stay on you while you stand still?'
Now ask children to run as fast as they can.
'Do you need to hold the paper for it to stay on you?'
What holds it there?

On a still day you are pushing against the air; on a windy day the air pushes against you.

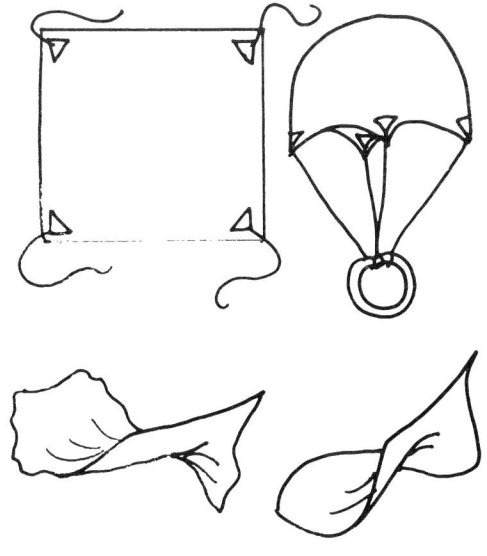

Pushing air against things

Try a pinwheel standing still — blow on it, then move the hand holding it very quickly. Now try the pinwheel running.
Try a kite on a calm day and on a windy day (see end of chapter).
Blow bubbles using wire loops, and bubble mix.
Watch bubbles float away in the air. Ask 'What are bubbles made of?'
   Air inside a skin of bubble mix.
   The air is holding up the bubbles.

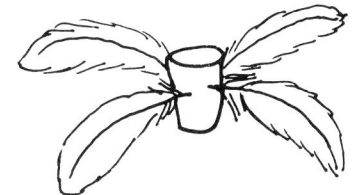

Strong bubble mixes are:

1. 1 tablespoon sugar in 3 tablespoons hot water, plus 9 T detergent
2. ¼ cup soap flakes in 1 cup hot water, left to gel for 3 days.

Make parachutes and spinning toys.

Drop these from a height — a climbing frame or a tree house.
Air can slow things falling as it pushes against them.

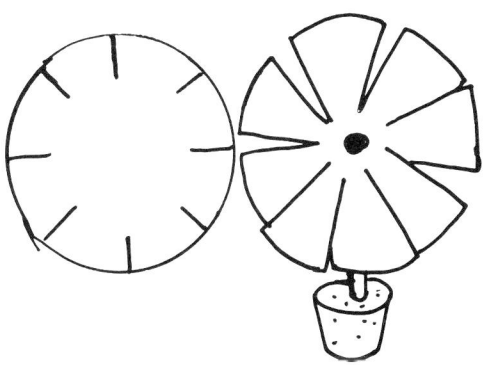

Pushing air against liquids

Punch two holes in the seal of a preserving jar. Screw the lid on to a jar full of water.

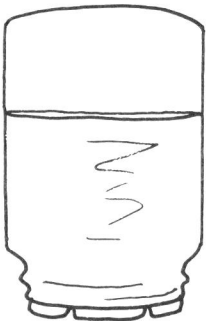

Quickly invert the jar full of water and hold it level. Very little water should drip out.
What holds the water in?

Air pressing all around the jar pushes up the water at the holes and stops it falling out.

Now hold the jar on a slope so that one hole is above the other.
What happens?

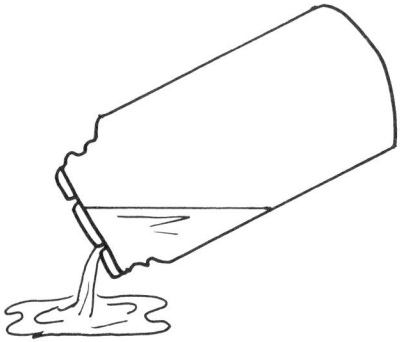

Air can now get into the top hole so that the air pressure is the same on both top and bottom of the water and the water can fall out of the bottom hole.
Put a finger over the top hole.
What happens?

Why do you have to punch *two* holes in a can to pour out the contents, e.g. condensed milk, tomato sauce?

Fill a straw with water and hold it vertically out of the water. What happens?
Water runs out.
Fill straw again; this time hold your finger over the top before you lift it out of the water.
Why does the water stay in the straw?
 Ask 'Where is the air pushing? Why can't the air push on the top of the water?

Warm air expands

Blowing up a balloon with a bottle

Materials:

A balloon and a plastic shampoo bottle.

Put balloon on bottle
Stand in hot water
Then stand in ice cubes

Ask: 'If we blew air into the balloon, what would happen?
 It would get bigger, it would expand. The air makes the balloon expand.

Where did the air come from?
 The bottle, because air is pushed into it. The air expands when it gets warm and takes up more space. It fills the bottle *and* some of the balloon.

Now put the bottle into a jug of ice-cubes. What happens?

Air contracts when it's cold and takes up less space. The balloon is pushed down into the bottle.

Bonus experience

A tetrahedral kite from milk straws

This requires dexterity and patience but the product is very strong and flies well. It is recommended for older or well coordinated children.

Materials:

Six milk straws
Some shirring elastic, or hat elastic
A packet of crepe paper
A sheet of tissue paper
Some glue suitable for sticking paper
A reel of button thread or thin fishing line
A sewing needle with a large eye
A rubber band
A pair of scissors
A ruler
A soft lead pencil

You may be surprised to know that straws can make a framework strong enough for a kite. It is the shape of the frame which gives it the strength *not* the materials from which it is made. Engineers use frames like this when they want to make a strong building out of lightweight materials. The shape of this kite frame is called a tetrahedron.

Making the frame

1. Cut a piece of shirring elastic *four* times the length of one straw.

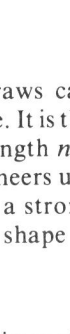

2. Use the needle to thread the elastic through three of the straws. Tie the ends of the elastic together so that the straws make a triangle as shown below.

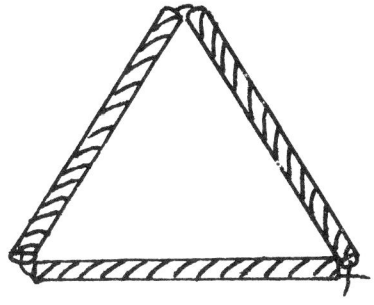

3. Cut three pieces of elastic *twice* as long as one straw. Tie them all together in a knot at one end. Using the remaining straws, thread one on to each piece of elastic.

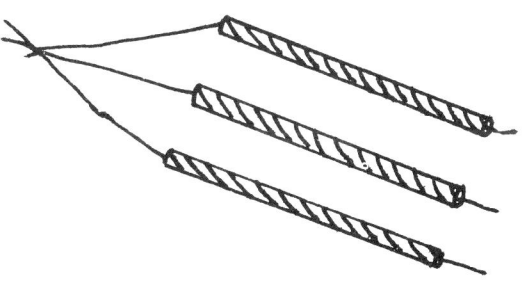

4. Tie the loose ends of the elastic to the corners of the first triangle by looping them round the elastic holding that triangle together. This makes the tetrahedron. The elastic must be stretched so that it is pulling the straws together, but it must not be too tight or it will bend them.

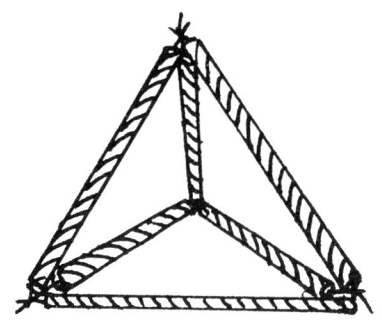

5. Cut away the loose ends of the elastic. The tetrahedral frame is now finished and is ready for covering.

Covering the frame

6. Take the sheet of tissue paper and fold it in half. Lay the tetrahedral frame on the folded sheet so that one of the straws is level with, and close to, the folded edge. Draw round the frame as shown below leaving a margin of about 2 centimetres.

8. When you open out the triangle you have cut, it will form a diamond shape. This has to be glued to *two* faces of the tetrahedron. Spread glue on the three straws of one triangular face (it does not matter which). Carefully press the glued face on to the tissue paper so that one straw is on the crease that was the folded edge. Bend the tissue paper upwards as in the picture below, and glue the second face in the same way as you did before.

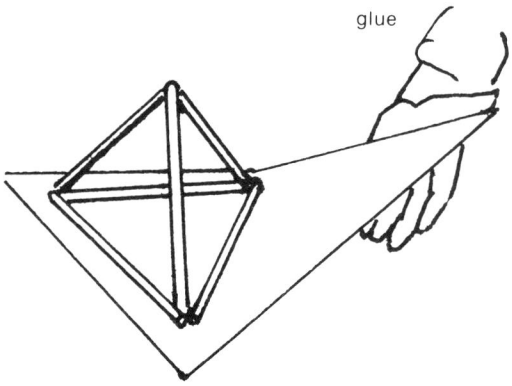

Fitting the tail and towing bridle

9. Trim the corners, then fold and glue the tissue paper to the inside.

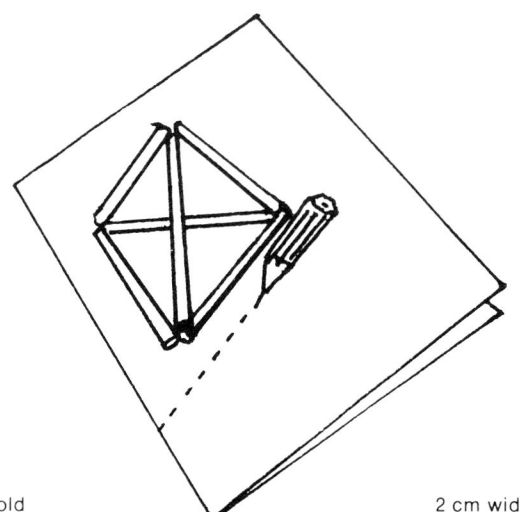

fold   2 cm wide

7. Cut out the triangular shape you have drawn but do *not* cut along the folded edge.

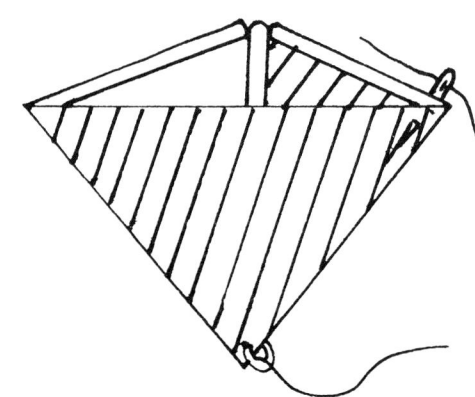

10. Tie a length of thread to each of the two corners of the kite covered in tissue. Use a needle to pass the thread through the tissue and round the straw, as close as possible to the corner, as shown above.

11. Bring the two threads together and tie a knot to form the bridle. The knot should be near the top corner of the kite, but you can experiment to find the best position when you fly your kite.

For a kite made from straws 21 centimetres long, the distance of the knot from each end are shown in the picture below.

12. Make the tail from two streamers of crepe paper, each 2 metres long and 3 centimetres wide. Glue these to the bottom corner of the kite as shown on the right.

The kite is now ready to fly.

Use button thread or thin fishing line to tow it.

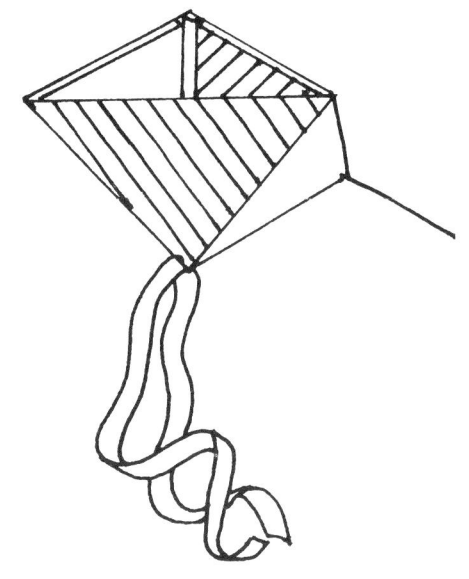

MAKING WORK EASIER

> *Basic Ideas*
>
> Things have weight.
>
> Simple machines such as ramps, rollers, wheels, gears and pulleys help us move heavy weights.
>
> Water and steam can also work for us.

CONTENTS OF THIS CHAPTER

Things have weight
Simple balances

Simple machines help us
Ramps
Rollers
Friction
Wheels
Axles
Gears
Pulleys

Water and steam can do work
Moving water
Steam

Things have weight

Most children today are familiar with terms such as astronauts, force-field, gravity-fields, and even anti-gravity.

Gravity is the pull of the earth on every body, that is, weight.

Play a 'what-if' game — What if everything was pulled up instead of down? When we got out of bed in the morning we'd be pulled up to the ceiling not down to the floor.

(In fact, we'd have to sleep under the bed).

If you threw a ball it would drift up out of reach.

To hold milk in a cup you'd need to hold the cup upside down, or milk would spill all over the ceiling.

Simple balances

These can be used in a sandpit,
                or the dough table,
                          at the water trough.
Drill a hole at the ½ metre mark of a metre rule. Loosely nail this to an upright nailed into a base. Children could help make this.

Place smaller nails at 5 cm intervals along the rule.

Make handles for empty yoghurt or margarine containers with wire or pipecleaners.

Compare equal weights at the ends of the scale.

Do they balance?
Now hang one container much nearer the middle than the other.
They are the same weights but they don't balance.
    Why?
Relate this to children on a seesaw. Two children of the same weight, one at each end of a seesaw balance it. If one child now moves closer to the fulcrum they are no longer in balance.

Two children of unequal weight, or adult and child, can balance the seesaw by sitting unequal distances from the fulcrum.

Simple machines help us

Ramps

Ask children to try lifting a suitcase with 5 kg weight in it onto the table.
 Ask 'Is it easy to lift the case?'

Lean a plank at least 1.2 metres long against the table.
 We call this a ramp.

Ask 'Can you slide the case up the ramp?' The ramp helps you do the work because it holds some of the weight.

Relate this to steep roads up hillsides.
 Ask 'Do the roads go straight up? Why do they slope gently up?' Try this in the sandpit.

In the block corner, encourage experiences with ramps of various slopes, comparing speed of vehicles and smaller blocks. Sit a block on a ramp.
 Ask 'How steep can I make the ramp before the block slides.?'

A screw is an inclined plane. Try a car jack to lift a full water bath.

Rollers

Supply boxes large enough for children to fit into. Ask 'Can you push each other along the concrete or floor?'

Now, introduce the inner cores of newsprint reels. Place the loaded box on these. *Supervise.*
Ask 'Do rollers make pushing easier?'

Talk about friction:

Feel the heat when the boxes are pushed along the ground. Friction always works against moving objects. Feel car tyres before and after a journey.
 Rub your hands together fast.
 Feel the heat as friction works on both hands.
 Try unscrewing the cap of a jar. 'You did that easily.'

Now make your hands soapy and try to unscrew the cap.

'You can't do it with slippery hands — you need *friction* to help do some work.'

Friction allows us to walk — when there is no friction, as on ice, we slip and slide.

Wheels

Wheels can be found everywhere in homes, schools, on streets and farms. They help us work.

Introduce axles.

Play with paper plates. Ask 'Can you roll them to each other? Do they move easily? Do they go straight?'

Find the exact centre of each plate by finding where two diameters cross, and cut out a ¼" hole.

Get children to join up two plates with straws and masking tape.

Do they roll better as a pair with an axle? Try positioning the axle off-centre.

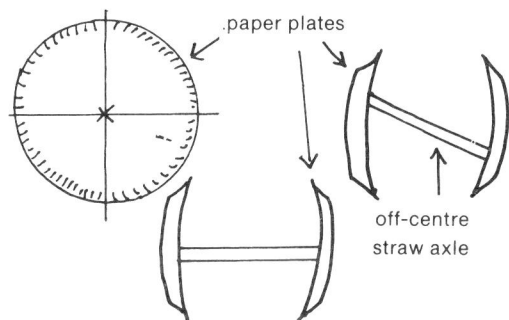

What happens? Investigate tricycles, doorknobs, typewriters, steering-wheels, rear wheels of vehicles and trailers.

Gears

These are wheels with teeth turning other wheels with teeth.

Look at eggbeaters.

Try beating water and detergent with a spoon and with an eggbeater — which works better?

Investigate old clocks, music boxes, bicycles and cog-wheel toys.

Pulleys

Pulleys are single or double wheels which help us lift and pull. Try lifting a bucket of sand.

Now attach a pulley above the bucket, thread rope through it, and try lifting the bucket by pulling on the rope.

Try with a bag of sand on a table.

Is it easier to pull the bag with or without the pulley?

Try a two-pulley system.

Water and steam can do work

Moving water can do work

Run water down a length of guttering to sail matchstick boats.

Use a jet of water from a plastic squeeze bottle to roll straws about.

Water turns a waterwheel and turbines (which run large generators to produce electricity).

Make a water-wheel using: a large cork,
    staples,
    2 nails,
    2 cotton reels,
    tap or hose with running water.

Push the staples into the sides of the cork and a nail into each end.

Slip each nail into the hole of a cottonreel, and with a hand holding each reel hold the staples under running water.

As water hits the staples it turns the cork.
Feel the *power* as it turns in your hands.

Steam can do work

Make a water-wheel as above.
Put water in a tin with a tight-fitting lid. Punch one hole in the lid and heat the water using a candle. A golden syrup tin stood on bricks works well.

Hold the water-wheel above the hole.
What happens?
What makes it turn this time?

HEAT

> *Basic Ideas*
>
> Fire needs fuel, air and heat.
>
> Some things burn, others don't.
>
> Heat makes things expand.
>
> When work is done, heat is formed as a by-product.
>
> Some things conduct heat, others don't.

CONTENTS OF THIS CHAPTER

Lighting fires

Fire needs fuel and air and heat
A beach volcano
Candles
Fire-extinguishers
Candle see-saw
Balloon bubbles in a bottle

Finding out what burns
A hobo stove
Flower pot heater
Sugar fire
Gas

Sorting burnables from non-burnables
Heat makes things expand
Buddel thermometer
Jammed glasses, tight lids
Heating milk
Heated smells

When work is done, heat is formed

Good and poor conductors
Underwater fire
Paper saucepan
Fire-guard

Lighting fires

Parents are often concerned about children playing with fire. Children who have been given the opportunity to learn about fire under adult supervision in safe and controlled ways are not likely to experiment secretly and dangerously.

Fire has a fascination for children of all ages. Family barbecues are unlikely to be suitable for children to learn about fire because they are monopolised by Dads. Also they are too hot, too high and too wobbly.

At preschool, place the barbecue bowl on bricks, or on the ground (lift the turf and later replace it). Involve children in the preparations including collecting worms from the dirt, filling buckets of water for safety precautions, collecting firewood, rolling newspapers to start the fire, scrubbing potatoes and wrapping them in tinfoil for cooking.

During this time there can be lots of talk about what is needed for fire (heat, fuel and air), why fire is dangerous, and what safety precautions are needed.

While at first they may require fairly authoritarian management, if a fire is a fairly regular part of the preschool programme children soon become more competent and help with lighting, peeling and putting in potatoes.

They all love helping to put the fire out!

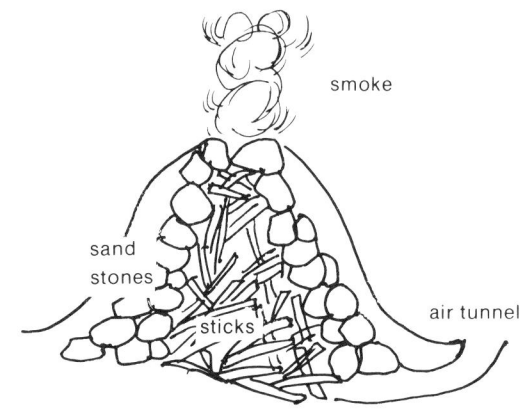

Fire needs fuel, air and heat

Collect sufficient light sticks to set two small fires.

Roll up newspaper for lighting both.

On one fire, arrange sticks and kindling flat and close together.

On the other leave plenty of spaces for air — a pyramid design is a good one.

Stack paper, sticks and dry seaweed for pops. Cover with sand, leaving an airhole tunnel at the base.

Leave the top of the cone open.

Light at the bottom through the airhole tunnel.

WARNING: Don't use stone unless you are certain they are shatterproof volcanic stones such as used in a hangi.

Candles

Stand a candle on a saucer and light it.

Give the children time to talk about their ideas of burning and heat.

Stress the danger of playing with matches.

Ask 'What is burning?'

Many children think wood is needed to make a flame. Here only the wick and wax are being burnt.

Invert a preserving jar over the candle.

The candle soon goes out because the oxygen in air gets used up by the flame.

Compare candles to show that burning uses air.

Light both fires at the same time. Which one burns best? Why? Both have fuel and heat. Discuss this together.

Beach volcano

This is another experience in fire making.

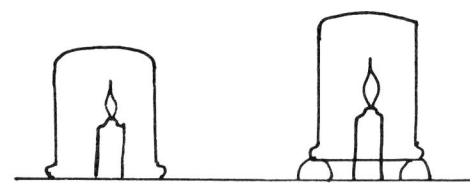

Invert jars over two candles of equal size.
Raise one jar above the table or saucer level with small pieces of plasticine or dough.
Which candle burns longest?

Burning uses air. Heat from the wick melts then vapourises the candle wax. This hot vapour burns, using oxygen in the air.

Try using three small candles, putting two of them under two different sized jars; which 'jar' lasts longer, the small or big one? Why does the candle with no jar over it keep burning?.

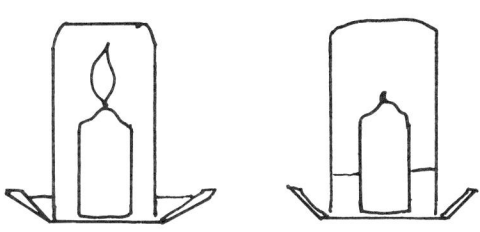

Add water to one saucer before you invert the jar over the burning candle. 'Why does water go into the jar?'

It replaces the burned oxygen; it is pushed into the vacuum by the air pressure outside the jar.

Fire-extinguishers

In an empty glass, light a candle stump.
In another glass mix 1 tsp. bicarbonate of soda (baking soda) and some vinegar and let it froth.

Tilt this glass over the candle.
It goes out. Why?

The chemical reaction of baking soda and vinegar formed carbon dioxide. This is heavier than air and doesn't burn. Therefore it falls on the flame and smothers it.

Many fire extinguishers work in this way. The sprayed foam is bubbles of carbon dioxide which blocks the oxygen supply, so the fire goes out.

A fire can be put out by:
- depriving it of fuel,
- depriving it of oxygen (CO_2 extinguisher),
- smothering in earth or sand.

Try all these methods of extinguishing fires.

Candle see-saw

Push a darning-needle through a cork sideways.
Attach a candle by its base to each end of the darning needle (the candles need to be the same length).
Push a knitting needle or a nail through the cork lengthways and suspend the cork between blocks or upturned glasses.
Light both candles.
They will rock vigorously.

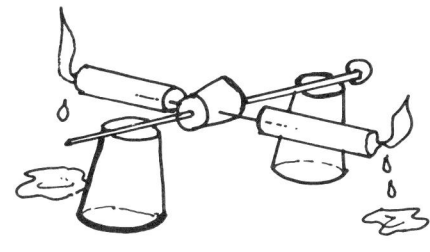

As a drop of wax falls from one end that end becomes lighter and goes up. Since the candles drip alternatively they go alternatively up and down. The fuel at each end is gradually being used up.

Balloon bubbles in a bottle

Throw a piece of burning paper into an empty milk bottle.

Stretch a piece of balloon rubber firmly over the mouth and hold it.

After a few moments the rubber is pushed into the neck of the bottle, and the flame goes out.

Why?

The burning paper used up oxygen in the bottle. This made a difference in air pressure between the outside and inside of the bottle; the higher outside pressure pushed the flexible balloon inwards to equalise pressure.

Finding out what burns

Fuels needed: wood, wax, methylated spirits, kerosene.
Use these in home-made stoves such as:

A hobo stove

Use a large, empty fruit juice can with one end removed.

With a pointed bottle or can opener, make holes around the top and bottom sides of the can.

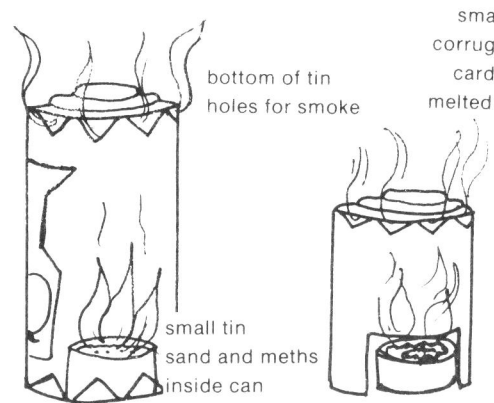

This is your hobo stove.

To heat it, either stuff the can with paper and twigs; *or* firmly wrap a half candle in corrugated cardboard and stuff this into a 450 gram can. Melt the remaining half candle, *or* use paraffin wax to soak the cardboard with wax; *or* half fill a small flat can with dry sand over which a few spoonsful of methylated spirits or kerosine **has been poured**. The sand should be damp but not soggy with fuel. This can be safely lit. Only the surface fluid ignites and its heat slowly turns the fluid below the sand to a burnable gas.

CAUTION: Keep bottles of fuel well out of children's reach. Use childproof caps if possible.

Light the burner in a sandpit and place the hobo stove, with the flat bottom uppermost, over the burner.

Butter this surface and use it to cook pikelets or omlettes, popcorn or to roast peanuts (move about to keep from burning). Even sausages and eggs can be cooked this way.

After use put a large can or metal bucket over the whole stove and push well into the sand. The fire will suffocate and the tins can be left to cool.

Flower pot heater

Light a hobo stove burner and place a dry unglazed flowerpot, supported a little off the ground, over the fire.

Heat radiates from the flowerpot and warms our hands and the room.

Sugar fire

Place a cube of sugar on a tin lid and try to light it. Now place a little cigarette ash on the lid and try to light it.

You won't succeed, either time.

Now dab a trace of cigarette ash on the cube of sugar and hold a burning match there.

The sugar burns with a blue flame until it is completely gone. Neither ash nor sugar will burn alone, but the ash initiates the burning of the sugar. We call the ash a catalyst.

Gas

Arrange a camping gas burner. Make tea, milo etc. or do some camp-cooking.

Sorting burnables from non-burnables

WARNING: Don't experiment with volatiles or plastics which may explode or give off poisonous gases.

Put match sticks, pieces of paper, milk bottle tops, straw, cloth and a few stones on a hobo stove, or on a big tin lid.

Heat strongly over camp stove.

Cool and examine.

Have any of the materials changed?
  Which have burned?
  Did any leave an ash?

Using the same materials as above try to light each one in turn with a match, or candle.

Which burn?

Is the statement made by a boy 'that things which have once lived will burn; things which have never lived won't' true?

Examine pieces of coal and coke. Note colour and hardness.
  How are they formed?
  What do they tell us?

Burn a match. Blow out the flame. Mark paper with the charred end. What is this? Make bigger pieces of charcoal by quenching partly burned firewood in a metal bucket of water.

Put very small pieces of wood shavings and cotton wool on tin lids. Try to light them. Discuss differences.

Clear your table. Pour a very little methylated spirits onto a tin lid. Light with a match. Note colour of flame.

Find out how a methylated spirits burner works.

Pour only a few drops of turps onto a piece of cloth. Put it on a tin lid and — standing well clear —light it.

Note colour of flames.

Heat makes things expand

A Buddel thermometer

Pour some coloured water into a bottle.
Drill a hole through the cork and push a straw through far enough that it dips into the water. Seal the cork with glue. Place your hands firmly around the bottle.

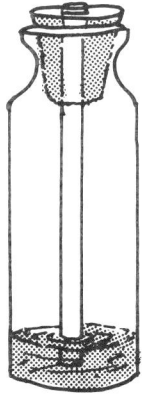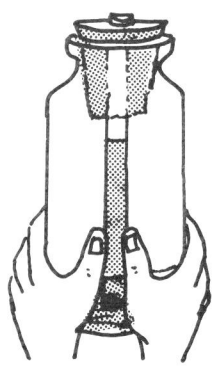

The water rises up the straw.

Your hands heat the air in the bottle which presses on the water surface. This pushes the water up the straw — the more the air is heated the further up the straw the water rises.

You could fix a scale on the side of the bottle.

Jammed glasses, tight lids

Fill the inner one with cold water, and stand the outer one in hot water.

Which expands? Which contracts?
Their sizes change, and they are easily parted.

Tighten a plastic coffee jar lid too tight for little hands to open. Hold the lid in hot water, and the heat soon expands the lid enough to be undone easily. The lid is now a little bigger than the jar. A metal lid on a sauce bottle can be used.

Heating milk

Put some milk in a saucepan, then heat it.

Heat expands the milk so much that gas bubbles form, and it froths up, expanding enough to overflow the saucepan.

What is the gas that forms?

Heated smells

Vanilla, cloves, onion, and orange skin. Can you smell them from the other end of the room? Now heat them. Can you now?

When work is done, heat is formed

Running makes you hot. You use up energy from food to do work (running, or digging, or playing) and in the process you get hot.

Electricity works to get through wire. The wire gets hot (and produces magnetism). This is how heaters work.

Light bulbs get hot, though their main function is to make light from electricity.

Friction makes heat too — feel a drill or car tyres or engines after use, or a nail yanked out of timber.

Good and poor conductors

Heat a metal knitting needle or stainless steel knife supported by cup handles. Feel where the heat is conducted to along the needle or knife. How close can you move your finger to the flame?

Now balance an old ceramic plate on a cup over the flame. 'How close to the flame can you move your finger. Take care!'

NOTE: Have a cup of cold water handy in case of overheated fingers. Good first aid education.

Underwater fire

Stick a candle stump in a bowl.

Fill the bowl up with water right to the rim of the candle.

Light the candle and watch it burn until the flame is *below* the level of the water.

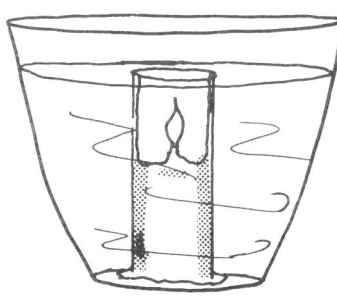

The flame will hollow out a deep funnel, leaving an extremely thin wall of wax which keeps the water out.

The water conducts so much heat from the candle that the outer layer of wax doesn't reach melting point. Therefore the wax cannot vapourise and burn. The water, of course, gets hot.

Paper saucepan

Boil water in a paper cup over a candle flame. Push a knitting needle through the top of a paper cup and suspend it between blocks or bottles.

Light a candle under the cup.

The water will boil, and the cup won't even be scorched.

'Why doesn't the cup burn?'

Because the water continually conducts heat away from the paper.

The paper never gets near the temperature it has to go before it can burn.

Fire-guard

Will flame burn through a sieve?

Light a candle and hold a metal kitchen sieve in the flame.

The flame does not go through it! Why?

The candle wax vapour goes through the wire but the wire mesh takes (conducts) nearly all the heat away from it. The vapour is not hot enough to burn.

This is why fireguards are made from wire mesh.

EARTH SCIENCE

Basic Ideas

Rocks are very old.

There are many kinds of rocks.

Rocks change and are worn away.

Soil consists of worn rocks and decayed plants and animals.

Land forms are shaped by the elements.

CONTENTS OF THIS CHAPTER

Rocks are very old
Fossils

Investigating rocks
Looking inside rocks

Erosion and sedimentation

Investigating soil

How land forms are shaped

Rocks are very old

Most children have some conception of long, long, ago in terms of dinosaurs. Dinosaurs lived millions of years before any people were on earth, and rocks can be even older than this.

Fossils

Show how fossils are made. A fossil is a print of a very old plant or animal, which filled up with minerals, and then set as hard rock.

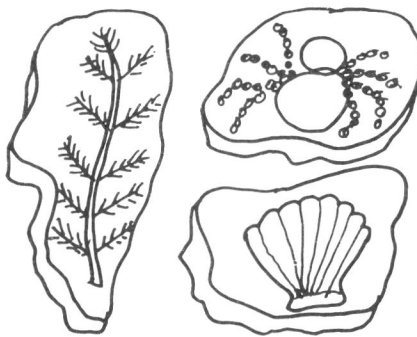

Roll out a slab of clay. Press a leaf into the clay and roll more clay on top.
 Now, if minerals could seep through the clay and set in the leaf, the leaf would become fossilised.
 Look at a fossil sample, if possible. Fossil shells are often found in soft sandstone and mudstones. Split such rocks along their grain using a cold chisel and hammer.

Investigating rocks

Collect a bucketful of rocks: road gravel, pumice, coal, chalk.
 Have children wash the rocks with detergent and water.
 Ask 'Do they look the same when dry and when wet?'

Sort rocks into various groups according to colour, shape, size, patterns, hard/soft (try scratching with a coin), light/heavy, rough/smooth, jagged, worn, with/without crystals.

If you and your children are interested, use a reference book to classify rocks (*igneous, sedimentary,* or *metamorphic*).

Looking inside rocks

Try to break the rocks with your hands.
If that is too hard, use a hammer. Place each rock in a sack to avoid flying splinters.
Ask 'What do you see inside the rock — is the colour the same as the outside? Do small bits of sand fall out?'

Use a hammer to pound each rock up into small pieces. Pound them as small as possible. To collect the bits put the rocks in a double plastic bag, or between layers of paper. Now mix the pounded rocks with a little water.
Ask 'Do you get a slimy mud?' 'What colour is the mud?' 'Are the mud or sand from different rocks different from each other?'
Sand and mud are made from rocks by the pounding of wind, rivers, frost, rain and sea.

Erosion and sedimentation

Use paint roller-trays or baking trays with sides will do.
Make a **sand castle**, a **clay castle**, and a **dirt castle**, at the top of the sloping trays.
Use tins with holes in the bottom to simulate rain.
Supply water from a gently-running hose.

Ask What happens to the castles when they are rained on? Do rivers run down the sides of the hills? Do they get deeper? Does the water run away with sand, clay and dirt? What happens at the bottom of the hills? Can you see the grains of sand etc. moving along with the water?

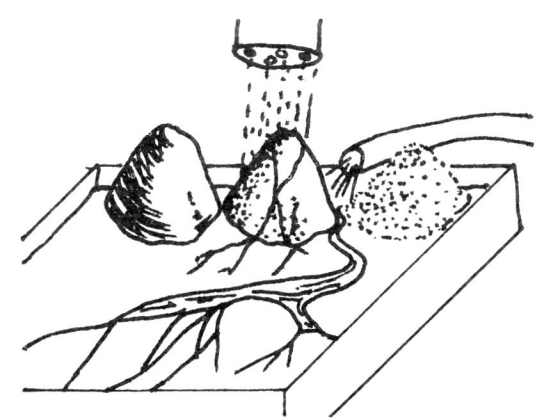

Try one tray very steep, another identical one not so steep.
Does more *erosion* occur on the steeper slope?

Investigating soil

Soil is made up of worn rocks, dead and living plants and animals, air and water.

Dig up some soil from the garden. Look at it closely. Sort it into piles:
- living things,
- things that were once alive but are now dead,
- things that were never alive.

Talk about what you have found in each pile.

Half fill a bottle with garden soil and then fill it with water.
Cover and shake well.
Watch the mixture settle. Ask 'Can you see the sediment dropping out of the water?'

Look for **road cuttings** where you see rusty crumbling weathered rocks turning into soil above freshly broken hard rock.
Dig a hole, trim a bank or follow drainlayers etc. to find layers in the soil. You may see soils, clay, pumice, sand and pebble layers, or find solid rock or ironpan below the surface.

Land forms are shaped

Look around your neighbourhood.

Explore the locality. Compare what you find when travelling with what you have back home. Talk about streams, slow muddy rivers, fast gravelly rivers, waterfalls, rapids, lakes and tidal creeks; mudflats and estuaries; beaches, sand dunes, cliffs and caves; plains, terraces, rolling hills, mountains and snowfields; valleys and volcanoes; limestone crags, caves.

Look in road cuttings for sandstone which was once beach, mudstone which was once estuary, gravel which was once riverbed, pumice which was once erupted, and lava which once flowed from volcanoes.

Look there for shellbeds, clays and soils, sloped and twisted layered rocks which show how the ground has bent and changed.

Look at eroded places, tussocklands, grasslands, scrublands, forest and farms.

Include *man-made land forms* for housing, cut-and-fill earthworks for highways, dams and airports, railway tunnels and bridges.

Help children think about how the land has become the shape we see, by the action of the sea, wind, rain, rivers, glaciers, earthquakes, volcanics, tectonic movement; and by man's efforts to make a comfortable place to live.

Encourage them also to consider the effects people have on the land in terms of both conservation and utilization.

Use water in the sandpit, and build miniature countrysides there.

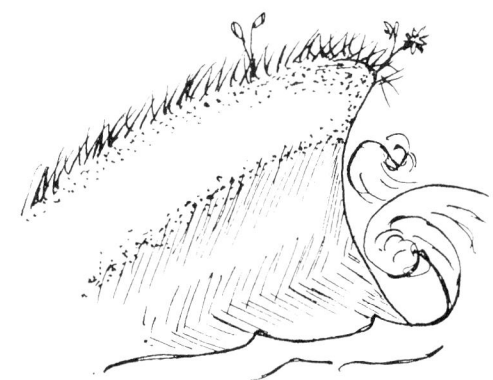

APPENDIX

Here is a brief classification of plants and animals which may help adults extend children's thinking and talking.

Children should be encouraged to look at, compare, think and discuss. There is little point in adults giving a lot of information unless it is done in a fun way. Adults don't even need to know much about classification to help children lay the foundations on which they will later learn to classify.

PLANTS
Plants which do not have flowers (they have spores).

1. **Single-celled plants** are similar to single-celled animals, but they are green and obtain their food differently. When in great numbers they often make pond-water look green. They are also common as a green powdery dust on tree trunks. The *algae* include the green slimy filaments on ponds. Seaweeds are also algae.
2. **Bacteria.**
3. **Fungi** include yeasts, moulds, toadstools, mushrooms, bracket fungi.
4. **Liverworts** are small flat, green, leaf-like plants found in clusters in damp places, stream banks, and in cellars and caves to which light has access.
5. **Mosses, lichens.**
6. **Ferns** are of many kinds and include bracken.
7. **Coniferous trees.** Examples are pines, kauri, miro, totara.

Flowering plants have roots, stems, leaves, flowers and fruits.

1. **Monocotyledons** are narrow-leaved plants with only one cotyledon in their seeds. Examples: grasses, reeds, rushes, cereals, iris, daffodil, cabbage tree.
2. **Dicotyledons** are broad-leaved plants with two cotyledons in their seeds.
(i) Herbaceous plants. Examples: daisy, buttercup, dandelion.
(ii) Shrubs. Wood, bushy plants. Examples: tea-tree, hebes, coprosmas.
(iii) Deciduous trees. Examples: oak, poplar, fruit trees.
(iv) Broadleaf evergreen trees. Examples: lacebark, lemonwood, lancewood, matipo, akeake.

ANIMALS

Animals without backbones: invertebrates

1. **Single-celled animals** are very abundant. These animals all live in water or moist environments and can usually be seen only with a microscope (amoeba etc.).
2. **Coelenterates** are the sea anemones and jelly-fish. Most live in the sea. They have hollow bodies.
3. **Flatworms** are mostly small fresh-water animals, often found under stones and floating leaves in streams. The group also includes the parasitic tape-worms and liver flukes. They are usually unsegmented.
4. **(True) Worms** include the earth-worm, and many little worms that live in ponds, and on sandy coasts. They are segmented.
5. **Crustacea** are crabs, lobsters, crayfish, shrimps and many small freshwater creatures such as the freshwater shrimp, water-flea and water-louse. They have a jointed exoskeleton, with a pair of limbs to each segment.
6. **Insects** (the biggest group) possess six legs and usually wings, and three parts to their bodies. Examples are butterflies, ants, bees, grasshoppers, flies, mosquitoes and beetles.
7. **Arachnids** have eight legs and no wings. There are web-spinning and hunting varieties of spider and mites in this group.
8. **Molluscs** are snails, slugs, whelks, oysters, and other "shell-fish", squid and octopus.
9. **Echinoderms** are the familiar starfish and sea urchin.

ANIMALS with vertebral columns: vertebrates.

A. 'Cold-blooded'

1. **Fish** breathe by means of gills and have bodies covered with scales. Examples: shark, herring, trout, bully. They live in water.
2. **Amphibia** are frogs, newts and toads. They have no scales on their bodies. They spend much of their lives on land, but their tadpoles live in water.
3. **Reptiles** are generally land-dwelling animals with scaly bodies. Examples: lizards, snakes, tortoises, crocodiles.

B. 'Warm-blooded'

4. **Birds** have bodies that are covered with feathers. Examples: sparrow, duck, hens, seagulls.
5. **Mammals** have bodies that are covered with fur; their young are born alive and suckled with milk. Examples: cows, dogs, cats, whales, seals, apes, man.

Acknowledgements

My sincere thanks to Playcentre parents and children in many parts of New Zealand who have shared science workshops with me. From these grew my inspiration to commit to paper the experiences which had all been found successful for children. Many Associations also sent me their written material and have continued to support my enthusiasm.

Special thanks to Jan Nalder, major contributor to chapters on living things; Alison Allen who argued, discussed, and tried out many experiments for herself; Alison Neale for her fire play contributions; Colin Walker, Science Adviser (Wellington Education Department) who checked scientific accuracy and made many helpful suggestions; Llyn Richards, and Caryl Hamer, editor, for continued support and technical expertise.

Laraine Peterson's art work has added a touch of humour while making the text clear. Her contribution to the book is invaluable.

My scientific understandings have been extended into all areas of daily living because of, and with, my own five investigating children. I thank them, and my husband John, for shared experiences; also for bearing with me as I've written, and for checking many experiments specifically for 'Mum's science book'.

Printed by Devon Colour Printers
7 Parkhead Place, North Harbour Industrial
Park, Albany, Auckland.